Lecture Notes in Control and Information Sciences

Edited by A. V. Balakrishnan and M. Thoma

For further listing of published volumes please turn over to inside of back cover.

Lecture Notes in Control and Information Sciences

Edited by A.V. Balakrishnan and M. Thoma

IFIP

36

Stochastic Differential Systems

Proceedings of the 3rd IFIP-WG 7/1 Working Conference
Visegrád, Hungary, Sept. 15–20, 1980

Edited by
M. Arató, D. Vermes, A. V. Balakrishnan

Springer-Verlag
Berlin Heidelberg GmbH 1981

Editors
M. Arató,
Research Institute for Applied Computer Sciences/SZÁMKI
P.O.B. 227
H-1536 Budapest (Hungary)

D. Vermes,
Bolyai Institute, University of Szeged
Aradi Vértanuk tere I
H-6720 Szeged (Hungary)

A. V. Balakrishnan,
Dept. of System Science, University of California
BH 4532 Los Angeles, CA 90024 (USA)

ISBN 978-3-540-11038-5 ISBN 978-3-540-38564-6 (eBook)
DOI 10.1007/978-3-540-38564-6

Originally published by Springer-Verlag Berlin Heidelberg New York in 1981

2061/3020-543210

P R E F A C E

This volume contains most of the contributions on the conference of stochastic differential equations at Visegrád /Hungary/ in September 198o. The conference was organized by IFIP TC 7 and the Hungarian Academy of Sciences. The support of Research Institute for Applied Computer Sciences /SZÁMKI/ was appreciated by the members of the conference. The main goal of the third conference in this area was to give a presentation of new results in stochastic system theory and its applications. The organizing committee wanted to stimulate the interest of more theoretical scientists working in this area in applications too, e.g. in engineering sciences or in computer science.

The papers in this volume cover many of the topics in stochastic systems. We hope that these papers give a good presentation of the present state of research in this field. We have to thank the authors for the careful preparation of their manuscripts. We hope that the ideas and methods presented in this book and at the conference would enforce the cooperation of scientists working on stochastic differential systems and in their applications.

The validity of stochastic system research in all over the world, the large number of interesting problems which remain to be solved, imply that we can look forward to the regular series of such conferences in this area.

The Editors

C O N T E N T S

ON OPTIMAL STOPPING TIMES IN OPERATING SYSTEMS

M. Arató

Research Institute for Applied Computer Sciences, Budapest

Introduction. In this paper we are concerning mainly with some general minimization /optimization/ problems of operating systems in large computer systems, which may be used in hierarchical memory allocation, page replacement, file allocation, routing and flow control, resource management problems, optimal load for disk devices, general tuning of files and I/O systems, workload characterizations.

The performance of a process executing in a computer system, e.g. with a memory hierarchy depends on both the memory referencing behavior of the process and the policy used to control the contents of each memory level. The direct observation and control of the referencing behavior involves a substantial overhead and radically perturbs the normal operation of the system, so cannot be applied in most cases. One of the important performance characteristic of program executing is the referencing behavior as determined by the program's reference string. In most cases it is possible to retain only a subset of the total address space of the process in main memory at any point of execution.
We formulate the problems in a probabilistic manner and try to solve them theoretically. Many of our models are not the usual in computer science, but it is possible to find the connections with simple mathematical statistical methods /Gomaa [8], Asztalos [4]/ and with queuing network methods too /Gelenbe [7], Kleinrock [9]/. The most important feature of such a treatment is the dynamic analysis of computer system performance, which is the only serious approach to the

design of performance control mechanisms /see e.g. Serazzi [1o], Arató [1]/.

Clearly, our results and their interpretations will be limited in the usual ways by the approach of mathematical modeling. That is, we shall use the analysis more as a mechanism for gaining insight into the system studied, rather than as a means for obtaining absolute measurements. Finally, interesting open problems related to our models develop when we remove the not necessary assumptions on Poisson arrivals and approximation with Wiener process.

Let we denote the execution of commands in a computer system by the following string $\xi_1, \underline{\xi_2}, \underset{\sim}{\xi_3}, \ldots, \xi_{p_1}, \xi_{p_1+1}, \ldots, \xi_{p_2}, \xi_{p_2+1}, \ldots,$ where e.g. underlying by $-$ means the commands of reference string of pages, underlying by $\sim$ means commands of interrupts. In the case when we are interested in the performance of a multiprogrammed system or a network system such questions arise as page fault rate, swapping rate, throughput, waiting time, response time and they all can be calculated from one or some of the different underlined sequences. This is the reason that instead of analyzing the whole system we consider a sub reference string e.g. $\eta_1 = \xi_2, \eta_2 = \xi_{p_1}, \eta_3 = \xi_{p_2+1}, \ldots$ the reference string of pages. From probabilistic point of view the sequence of random variables η_t $(t = 1,2,\ldots)$ may form an independent, identically distributed, or Markov, or LRU stack model, but in most cases with unknown probability distribution. The distribution has to be estimated in the course of the execution of the programs. In the sequel we shall investigate always only a subsequence of the executions as reference string /e.g. references of pages, references of files, .../.

An operating system handles all the system programs and user programs and there may be included also software monitors which collect the needed measurements for different reference strings.

However the dynamical handling /with software monitors/ involves a large overhead which manifests itself either as delays in execution or large storage requirements, or both. The execution delay results from the additional recording code which must be executed at the end of every traced instruction.

Recently, because of the storage requirement of large data-base systems, more attention has been given to mass storage systems, where depending on the system we may have different access mechanisms. Consequently the problem of minimizing the expected read/write movement /often we say head movement/ in one or more dimensional case has an important role. In addition, the head movement may be measured in different metrics.

1. Two and many armed bandit type problems

Now let us formulate some problem for independent reference string model, where

$$P\,\{\xi_t = i\} = p_i \qquad (i = 1,2,\ldots,n)$$

and distribution $\{p_i\}$ is unknown.

Example 1./Page fault/ Let k mean a fix number and D_t $(|D_t| = k)$ the set of those pages which are on the first level of memory hierarchy at time t and

$$\eta_t = \begin{cases} 1 & \text{if} \quad \xi_t \bar{\epsilon} D_t, \qquad \text{/page fault occurs/}, \\ 0 & \text{if} \quad \xi_t \epsilon D_t. \end{cases}$$

We want to minimize the expected number of page faults

$$\min_D E\,\left(\frac{1}{T}\sum_1^T \eta_t\right) = \min_D \frac{1}{T}\sum_{t=1}^T P\,(\eta_t = 1) = \min_D \frac{1}{T}\sum_{t=1}^T P\,\{\xi_t \bar{\epsilon} D_t\}.$$

Example 2./Optimal arrangement on linear storage/ Let $d(i,j)$ mean the distance of places i and j $(i,j = 1,2,\ldots,n)$ and $\pi(\cdot)$ an arrangement of records (the i-th record is on place $\pi(i)$). The optimization problem is the following

$$\min_{\pi} E\left[\frac{1}{T}\sum_{t=1}^{T} d(\pi(\xi_t),\pi(\xi_{t-1}))\right] = \min_{\pi} \frac{1}{T}\sum_{t}\sum_{i,j} p_i p_j \, d(\pi_t(i),\pi_{t-1}(j)).$$

Example 3./Dynamic file assignement/ Let us assume that a file may be allocated in one of the n computers and let $Y_i(t) = 1$ if at time t it is the i-th computer /for all others $Y_j(t) = 0$). Let $\eta_j(t) = 1$ if at time t the j-th computer requests the file, $P\{\eta_j(t) = 1\} = p_j$. If the transition cost of the file from one computer to another is 1 we want to minimize

$$\min_{Y} E\left(\frac{1}{T}\sum_{t=1}^{T}\sum_{\substack{i,j\\ i\neq j}} \eta_j(t)Y_i(t)\right) = \min_{Y} \frac{1}{T}\sum_{t=1}^{T}\sum_{\substack{i,j\\ i\neq j}} Y_i(t)\, P\{\eta_j(t) = 1\}$$

All in the above examples if $\{p_i\}$ are known the solution is almost trivial (see Arató [1] , Wong [12]). There may be investigated also the case when $\xi(t)$ forms a Markov chain, in this case the problem can be handled by dynamic programming (see Wong [12]). The case when $\{p_i\}$ are unknown, we are using the Bayesian method and the separation principle (see Wonham [13], for linear Gaussian processes) , which asserts that the maximum likelihood estimates of unknown probability distribution is sufficient for optimization. (Arató, Benczúr [3]).

On decision algorithms D_t, π_t, Y_t $(t = 0, 1, 2, \ldots)$ there have been restrictions as in other case the cost cannot be described in such simple form. D_t has to be a demand page replacement algorithm, π_t cannot be a rearrangement at every step, Y_t has to be also a demand dependent assignement.

The limit behavior (when $T \to \infty$) of the cost function and decision algorithm, is special application of the law of large numbers (see e.g. Wong [12], Arató-Benczúr [3]). We formulate it in the following form

Theorem 1. Let $f_j(t)$ $(j = 1,2, \ldots, N)$ denote the frequencies of event $\xi_t = j$ in an independent reference string model. Let D_t mean a restricted decision, which depends only on ξ_t, and has the cost $R(\xi_1 \ldots, \xi_t)$ at every step. If the mean value function on distribution $\{p_i\}$ is minimal at D^* :

$$\min_D E_{\{p_i\}}\left[\frac{1}{T}\sum_{t=1}^{T} R_D(\xi_1,\ldots,\xi_t)\right] = E_{\{p_i\},D^*}\left[\frac{1}{T}\sum_{t=1}^{T} R_{D^*}(\xi_1,\ldots,\xi_t)\right]$$

then

$$\lim_{T\to\infty} \frac{1}{T}\sum_{t=1}^{T}\left(\sum_{i_1..i_N} R_D(\xi_1,\ldots\xi_t)\cdot\frac{f_1(t)\ldots f_N(t)}{N^t}\right) \to E_{\{p_i\},D}\, R_D(\xi_t).$$

The two armed bandit problem type solution is true for every fixed T and under hypothesis that the apriori distribution is uniform. We can formulate it in the following way (the myopic policy)

Theorem 2. If at time moment t_o the frequencies of $\xi_t = j$ are $f_j(t_o)$ $(j = 1,2,\ldots, N)$ and with fixed probability distribution $\{p_j\}$ the optimal decision policy until time $t_o + T$ is $D*$ when $R(\xi_1,\ldots,\xi_t)$ is the cost function, i.e.

$$\min_D E_{\{p_i\}}\left[\frac{1}{T}\sum_{t_o+1}^{t_o+T} R_D(\xi_1,\ldots,\xi_t)\right] = E_{\{p_i\},D^*}\left[\frac{1}{T}\sum_{t_o+1}^{t_o+T} R_{D^*}(\xi_1,\ldots,\xi_t)\right]$$

then under the Bayesian approach with uniform distribution for the apriory arrangement of $\{p_i\}$ values on $\{i\}$ the optimal solution of minimizing the cost is the same as with known distribution:

$$\min_D E_{\{p_i\},\ \text{uniform}}\left[\frac{1}{T}\sum_{t=t_o+1}^{t_o+T} R_D(\xi_1,\ldots,\xi_t)\right]=E_{\{f_i\},D^*}\left[\frac{1}{T}\sum_{t_o+1}^{t_o+T} R_{D^*}\right].$$

Between the two-armed bandit problem and reference string models there exists a great difference, which causes that our statements have to be true without Bayesian assumption, i.e. the separation principle works. The difference consists of that the reference string ξ_t gives for probability distribution statistics, but in two-armed bandit problem without apriori distribution we cannot get it /in other words in reference string models ξ_t does not depend on decisions/.

2. Disorder type problems

The purpose of this section is to show that stopping rule procedures are needed in computer performance analysis. In many computer and network system problems arises the problem to detect changes in the behavior of the system.

In the following we study two examples

Example 4./Swapping processes/ In virtual storage operating systems the multiprogramming level, η, depends on the total storage demand of the programs, i.e. on the number of allocated pages, k_t, in main storage, at time moment t. A swap out decision, $\eta \to \eta - 1$ transition has to be made when k_t reaches some treshold, $k_t \geq M$, and a swap in decision, $\eta \to \eta + 1$ transition, if $k_t \leq L$.

If the treshold margin is small the overflow and underflow events could occur quite frequently and the performance effectivity is bad in the sense that the amount of overhead is great. If the margin is large storage may not be utilized effectively in the sense, that progress-rate, i.e. proportion of time that the CPU is in problem state, is small.

We assume that the swapping process in normal case is a Poisson process with rate λ_o which may change to λ_1 $(\lambda_1 > \lambda_o > o)$ at a certain random time ξ, or to λ_2 $(_o < \lambda_2 < \lambda_o)$ also at a certain random time. From state λ_1 or λ_2 the process may return to λ_o. Even in the case when there may be only one transition from λ_o to λ_1 /and back/ the situation is different from "classical disorder" problem. We have not the possibility to stop the observation and to verify that a change takes place or not. Let $\xi_1, \xi_2, \ldots$ mean the transitions from state λ_o to λ_1 or back and $\tau_1, \tau_2, \ldots$ the Markov moments when transitions were observed. Let $\chi_s = 1$ if the observed and "wanted" processes are not in the same state /see Fig.1./

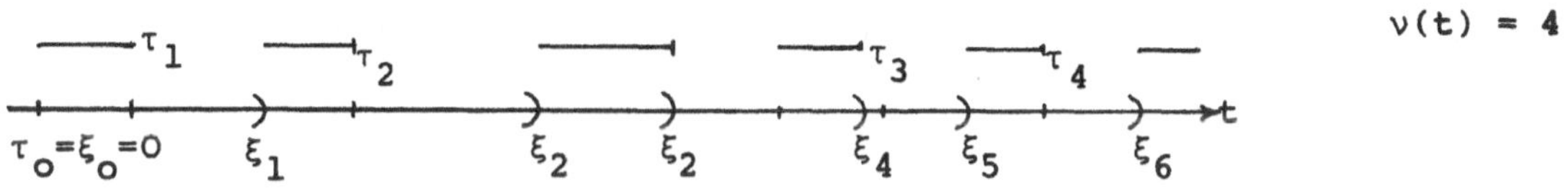

Fig.1.

We have to minimize $E \int_o^t \chi_s ds = \min$, under the condition that the number $\nu(t)$ of τ moments is not great $(E\nu(t) \le \lambda t$, where $P(\xi = o) = 1/2$, $P(\xi > t \mid \xi > o) = e^{-\lambda t})$.

Example 5 /Failure processes in data transaction systems/

/see [2] , [5] , [6] , [7] /.

The state of a data basis is assumed to operate in the checkpoint rollback recovery mode. A random time process between arrivals of transactions, also the time necessary for their execution /service/ is generally random.

Let b_i denote the i-th transaction time and c_j the j-th failure time $(b_o = c_o = o)$, and we assume that they are independent Poisson processes with rate λ and γ respectively. At specific instants of time a_i checkpoints are instaured /dump/; that is a secure and valid copy of all the information contained in the system is stored into a memory unit. The creation of the i-th checkpoint immobilizes the system durimg M_i time, which is a failureless run time necessary for a dump. N_i means the run time necessary for a load re-generating a control point after failure.
Further we assume that in normal processing the service time for each transaction has distribution $F_o(t)$ with $\frac{1}{f_o} = \int_o^\infty t\, dF_o(t)$, and after a failure the distribution is $F_1(t)$ with $\frac{1}{f_1} = \int_o^\infty t\, dF_1(t) > \frac{1}{f_o}$.

In other cases the service times of transactions after a failure have monoton increasing character i.e. for their distribution functions

$$F_1(t) \geq F_2(t) \geq F_3(t) \geq \ldots$$

/from this follows that $\frac{1}{f_1} \leq \frac{1}{f_2} \leq \ldots \leq \frac{1}{f_i} \leq \ldots$/.

Let the random variable C_1 mean the moment of the first failure with distribution

$$(2.1) \qquad P(C_1) = 0 = \pi, \quad P(T_o < t \,|\, T_o > 0) = 1-e^{-\gamma t}.$$

Let τ mean the stopping time where we stop the execution process to verify that there was a failure or not. The restriction that $\tau < a_1$ /or $\tau < a_{i+1} - a_i$ / must hold.
The observed process with σ-algebra $\mathcal{F}_t$ may be the process of service to request for transactions /case A/ or only the number of transactions waiting to be processed, including the one being processed /case B/.

Under the above assumptions we consider the cost function

(2.2) $$\rho^{\Pi}(\tau) = P^{\Pi}(c_1 > \tau) + c \cdot E^{\Pi}(\tau - c_1 | \tau > c_1)\, P^{\Pi}(\tau \geq c_1).$$

Parameter c depends on the sequence a_i /i = 1,2,.../, i.e. on the checkpointing times.

Without checkpoints, and in case A, when $F_1(t) = F_2(t) = \ldots$ we have to investigate the ordinary disorder problem /see Shiryaev [11] /, but with the modification that the number of observed random variables has Poisson distribution.

Now let us formulate the model in this simplest case. Let ξ_t mean the following process: $\{\eta(t), \xi_1, \xi_2, \ldots, \xi_{\nu(t)-m(t)}, \nu(t), m(t)\}$, where $\eta(t)$ is a Poisson process with rate λ, $\nu(t)$ is the number of requests until t $(P(\nu(t) = k) = \frac{(\lambda t)^k}{k!} e^{-\lambda t})$, $m(t)$ is the number of transactions waiting to be processed at time t, ξ_i is the service time of the i-th transaction. The observed process may be ξ_t or only a part of it, e.g. $m(t)$. The disorder /failure/ takes place as it is described in /2.1/. Assuming that the random variables $\xi_1, \xi_2, \ldots, \xi_{\nu(c_1)-m(c_1)}$ have distribution $F_o(t)$ and $\xi_{\nu(c_1)-m(c_1)+1}, \ldots$ have distribution $F_1(t)$ we want to minimize (2.2). Let

(2.3) $$\pi_t = P\{c_1 \leq t | \mathcal{F}_t^{\xi_1, \ldots, \xi_{\nu(t)-m(t)}}\}$$

and

(2.4) $$\tau^* = \inf_t \{t > 0 : \pi_t \geq A^*\}$$

for a fixed A^*.

We state the following theorem /see Shiryaev [11] Th.7. Ch.4.3/

<u>Theorem</u>. Let $c > 0$, $\lambda > 0$ and $\pi_0^\pi = \pi$ then the Markov moment τ_π^* defined in (2.4), is π-Bayesian

$$\rho^\Pi(\tau_\pi^*) = \inf_\tau \rho^\Pi(\tau),$$

where A^* can be constructed in the same way as in disorder problem in discrete time case (see [11] formula (4.129)).

R E F E R E N C E S

[1] M.Arató — Statistical Sequential Methods for Utilization in Performance Evaluation 287-3o3 /in Measuring, Modeling and Evaluating Computer Systems, North Holland /1977/, H.Beilner, E.Gelenbe/

[2] M.Arató — On failure processes in computer systems /in print/

[3] M.Arató - A.Benczúr — Dynamic placement of records and the classical occupancy problem, Comp. and Mths. with Applications 7/1981 /,173-185.

[4] D.Asztalos — A hybrid simulation / analytical model of a batch computer system 149-16o /in Performance of Computer Systems, North Holland /1979//

[5] A.Benczúr - A.Krámli — An example for an adaptive control method providing data base integrity, 263-276 /in Performance of Computer Systems, North Holland /1979/, M.Arató, A.Butrimenko, E.Gelenbe/

[6] K.Chandy - I.Brown - C.Dissly,- W.Uhring — Analytical models for roll back and recovery strategies in data base systems, IEEE Trans. on Software Eng.1 /1975/ loo-llo.

[7] E.Gelenbe,- D.Derochette — Maximum load and service delays in a data base system with recovery from failures 129-142 /in Modelling and Performance Evaluation, North Holland, /1977//

[8] H.Gomaa A modelling approach to the evaluation of computer system performance 171-2oo /in Modelling and Performance Evaluation, North Holland, /1977//

[9] L.Kleinrock Queuing networks, Vol.I. John Wiley, /1975/

[10] G.Serassi The dynamic behavior of computer systems, p.127-163 /in Experimental Computer Performance Evaluation, North Holland, 198o, Eds.D.Ferrari, M.Spadoni/

[11] A.Shiryaev Statistical sequential analysis, Nauka, /1976/, Moscow /in Russian/

[12] C.K.Wong Minimizing expected head movement in one-dimensional and two-dimensional mass storage systems. /Computing Surveys 12 /198o/ N°.2./

[13] W.Wonham On the separation theorem of stochastic control SIAM Journal, Control 6 /1968/ 312-326.

SEMIMARTINGALES DEFINED ON MARKOV PROCESSES*

by

E. ÇINLAR and J. JACOD

Our objective is to represent additive semimartingales defined over Markov processes in terms of well-understood objects like Wiener processes and Poisson random measures. In particular, if the underlying Markov process itself is a semimartingale, our results yield a representation for it in terms of stochastic integral equations driven by a Wiener process and a Poisson random measure. Thus, this work is in the tradition of research devoted to clarifying the structure of Markov processes by relating them to Wiener and Poisson processes. Previous results of this nature were obtained by ITO [10], FELLER [6], [7], [8], DYNKIN [4], SKOROKHOD [14], [15], and KNIGHT [13], just to name a few.

Our aim here is to give a semi-formal summary of the main results of [2], which is in turn a fairly long and careful rearrangement of results that have been available under different setups. Among the latter, the works of GRIGELIONIS [9], EL KAROUI and LEPELTIER [5], and JACOD [12] figure prominently in all this.

PRELIMINARIES

Let E be a topological space whose Borel σ-field $\underline{\underline{E}}$ is separable. Let $X = (\Omega, \underline{\underline{F}}_\infty, \underline{\underline{F}}_t, \theta_t, X_t, P_x)$ be a normal strong Markov process with state space $(E, \underline{\underline{E}})$, infinite lifetime, and right-continuous left-hand-limited paths. Here $\underline{\underline{F}}_t$ is the usual completion of $\underline{\underline{F}}^o_{t+}$ where $\underline{\underline{F}}^o_t = \sigma\{X_s : s \leq t\}$. See [1] for the precise meanings.

*Research supported by the Air Force Office of Scientific Research through their Grant No. AFOSR-80-0252.

We need to work with extensions of X and need larger filtrations than $(\underline{\underline{F}}_t)$. Let $(\underline{\underline{H}}_t^\circ)$ be a filtration on Ω such that $\underline{\underline{H}}_t^\circ \supset \underline{\underline{F}}_t^\circ$ and set $\underline{\underline{H}}_\infty^\circ = \vee_t \underline{\underline{H}}_t^\circ$. We assume that each $\underline{\underline{H}}_t^\circ$ is separable and that $P_x(d\omega)$ is a transition probability from $(E,\underline{\underline{E}})$ into $(\Omega,\underline{\underline{H}}_\infty^\circ)$. We let $\underline{\underline{H}}_t$ be the usual completion of $\underline{\underline{H}}_{t+}^\circ$, set $H_\infty = \vee_t \underline{\underline{H}}_t$, and let $\underline{\underline{H}} = (\underline{\underline{H}}_t)$. Then, $\underline{\underline{H}}$ is said to be a <u>strong Markov filtration</u> for X if for every finite stopping time T of $(\underline{\underline{H}}_{t+}^\circ)$ the following hold: $X_T \in \underline{\underline{H}}_T/\underline{\underline{E}}$, $\theta_T \in \underline{\underline{H}}_\infty/\underline{\underline{H}}_\infty^\circ$; $\underline{\underline{H}}_{(T+u)+}^\circ = \underline{\underline{H}}_{T+}^\circ \vee \theta_T^{-1}\underline{\underline{H}}_{u+}^\circ$ for all $u \geq 0$; and

$$(1) \qquad E_\mu[\, Z\circ\theta_T \mid \underline{\underline{H}}_{T+}^\circ \,] = E_{X_T}[\, Z \,]$$

for all finite measures μ on $\underline{\underline{E}}$ and all $Z \in b\underline{\underline{H}}_\infty^\circ$. Then, $(\Omega,\underline{\underline{H}}_\infty,\underline{\underline{H}}_t,\theta_t,X_t,P_x)$ is a strong Markov process in the sense of [1], but with the additional property that (1) holds for all $Z \in b\underline{\underline{H}}_\infty^\circ$ instead of holding only for $Z \in b\underline{\underline{F}}_\infty^\circ$.

From here on, $(X,\underline{\underline{H}})$ will always denote the Markov process X together with a strong Markov filtration $\underline{\underline{H}} = (\underline{\underline{H}}_t)$ for it.

<u>Semimartingales</u>. An additive semimartingale over $(X,\underline{\underline{H}})$ is a real-valued process Y that is adapted to $\underline{\underline{H}}$, is additive with respect to (θ_t), and is a semimartingale over the stochastic base $(\Omega,\underline{\underline{H}}_\infty,\underline{\underline{H}}_t,P_x)$ for every x. An m-dimensional semimartingale is an m-dimensional vector valued process whose every component is a semimartingale.

Let Y be an m-dimensional additive semimartingale over $(X,\underline{\underline{H}})$. We define Y_t^e to be the sum of the jumps of Y during $(0,t]$ whose magnitudes exceed 1; then, Y^e is an additive pure jump process adapted to $\underline{\underline{H}}$ and with only finitely many jumps over finite intervals. Now, $Y - Y^e$ is a semimartingale with bounded jumps, and therefore, can be written as the sum of a process Y^b of locally bounded variation and a process M that is a local martingale. Further, M can be decomposed into a continuous local martingale Y^c and a purely discontinuous (a compensated sum of jumps) local martingale Y^d. Hence,

(2) $$Y = Y^b + Y^c + Y^d + Y^e,$$

where each term is $\underline{\underline{H}}$-adapted and additive. Normally, this decomposition is unique up to a null set for the probability measure being employed, but it can be shown that (see [3]) it is possible to take the exceptional set to be the same for all measures P_x, $x \in E$.

We let $B = Y^b$, let $C = (C^{ij})_{i,j \leq m}$ where C^{ij} is the quadratic covariation of the ith and j^{th} components of Y^c, and let G be the dual predictable projection of the jump measure Γ of Y defined by

(3) $$\Gamma(B) = \sum_{s>0} 1_B(s, Y_s - Y_{s-})\, I_{\{Y_{s-} \neq Y_s\}}, \qquad B \in \underline{\underline{B}}_+ \otimes \underline{\underline{R}}^m.$$

See [12] for the precise meanings. Then, (B,C,G) is called the <u>triplet of local characteristics</u> for Y. As with the decomposition (2), (B,C,G) is defined up to a set that is null under every P_x. That this is possible was basically proved in [3] along with the following.

(4) THEOREM. Let Y be an m-dimensional additive semimartingale over $(X,\underline{\underline{H}})$, and suppose it is quasi-left-continuous (that is, $Y_{T_n} \to Y_T$ almost surely for every increasing sequence (T_n) of $\underline{\underline{H}}$-stopping times with finite limit T). Let ν be an arbitrary positive σ-finite infinite diffuse measure on $\underline{\underline{R}}$. Then, there exist

a) an $(\underline{\underline{F}}^o_{t+})$-adapted continuous strictly increasing additive functional A with $\lim_{t\to\infty} A_t = \infty$ and $dt << dA_t$,

b) an $\underline{\underline{E}}$-measurable bounded $\mathbb{R}^m$-valued function b,

c) an $\underline{\underline{E}}$-measurable bounded $m \times m$ lower triangular matrix-valued function $c = (c_{ij})_{i,j\leq m}$ with $c_{ij} = 0$ whenever $c_{jj} = 0$, and

d) an $\underline{\underline{E}} \otimes \underline{\underline{R}}$-measurable $\mathbb{R}^m$-valued function k such that

$$B_t = \int_0^t b(X_s)\, dA_s, \qquad C_t = \int_0^t cc'(X_s)\, dA_s,$$

$$G([0,t]\times B) = \int_0^t dA_s \int_{\mathbb{R}} \nu(dz)\, 1_B \circ k(X_s,z), \quad B \subset \mathbb{R}^m \setminus \{0\}, \text{ Borel},$$

define a version of the triplet (B,C,G) of local characteristics of Y under every P_x (here, c' is the transpose of the matrix c).

Markov extensions. Let $(\Omega',\underline{\underline{H}}',P')$ be an auxiliary probability space, set $(\tilde{\Omega},\tilde{\underline{\underline{H}}}_\infty,\tilde{P}_x) = (\Omega,\underline{\underline{H}}_\infty,P_x) \times (\Omega',\underline{\underline{H}}',P')$, and let π be the projection mapping $(\omega,\omega') \to \omega$ from $\tilde{\Omega}$ into Ω. For any random variable Z defined on Ω or Ω', we let the same letter Z denote its natural extension onto $\tilde{\Omega}$, e.g. if Z is defined on Ω, then $Z(\omega,\omega') = Z(\omega)$. Let $\tilde{\underline{\underline{H}}} = (\tilde{\underline{\underline{H}}}_t)$ be a right continuous filtration on $(\tilde{\Omega},\tilde{\underline{\underline{H}}}_\infty)$, and let $(\tilde{\theta}_t)$ be a semi-group of transformations on $\tilde{\Omega}$. Set $\tilde{X} = (\tilde{\Omega},\tilde{\underline{\underline{H}}}_\infty,\tilde{\underline{\underline{H}}}_t,\tilde{\theta}_t,X_t,\tilde{P}_x)$. The pair $(\tilde{X},\tilde{\underline{\underline{H}}})$ is said to be a strong Markov extension of $(X,\underline{\underline{H}})$ if $\pi \in \tilde{\underline{\underline{H}}}_t/\underline{\underline{H}}_t$ and $\pi\circ\tilde{\theta}_t = \theta_t\circ\pi$ for all $t \geq 0$ and if for every $Z \in b\tilde{\underline{\underline{H}}}_\infty$ and finite $\tilde{\underline{\underline{H}}}$-stopping time T, $Z\circ\theta_T$ is measurable relative to the completion of $\tilde{\underline{\underline{H}}}_\infty$ and

$$\tilde{E}_x[\, Z\circ\tilde{\theta}_T \mid \tilde{\underline{\underline{H}}}_T\,] = \tilde{E}_{X_T}[\, Z]].$$

Wiener processes. A wiener process over $(\tilde{X},\tilde{\underline{\underline{H}}})$ is an m-dimensional continuous additive process $\tilde{W} = (\tilde{W}^i)_{i\leq m}$ adapted to $\tilde{\underline{\underline{H}}}$ such that the increments $\tilde{W}^i_u\circ\tilde{\theta}_t = \tilde{W}^i_{t+u} - \tilde{W}^i_t$ are independent of each other and of $\tilde{\underline{\underline{H}}}_t$ and have Gaussian distributions with means 0 and variances u under every $\tilde{P}_x$, $x \in E$.

Poisson random measures. A random measure $\tilde{N}$ on $\mathbb{R}_+ \times \mathbb{R}$ over $(\tilde{X},\tilde{\underline{\underline{H}}})$ is a positive transition kernel from $(\tilde{\Omega},\tilde{\underline{\underline{H}}}_\infty)$ into $(\mathbb{R}_+\times\mathbb{R}, \underline{\underline{\mathbb{R}}}\otimes\underline{\underline{\mathbb{R}}})$. It is said to be additive if $t \to \tilde{N}([0,t]\times B)$ is an $(\tilde{\theta}_t)$-additive process for every $B \in \underline{\underline{\mathbb{R}}}$. An additive integer-valued random measure $\tilde{N}$ over $(\tilde{X},\tilde{\underline{\underline{H}}})$

is said to be a Poisson random measure on $\mathbb{R}_+ \times \mathbb{R}$ with mean measure dt $\nu(dz)$ if for every collection $\{B_1,\dots,B_j\}$ of disjoint Borel subsets of $(t,\infty) \times \mathbb{R}$ the random variables $\tilde{N}(B_1),\dots,\tilde{N}(B_j)$ are independent of each other and of $\tilde{\underline{H}}_t$ and are Poisson distributed with means $n(B_1),\dots,$ $n(B_j)$ where $n(B) = \int_B dt\ \nu(dz)$.

If $\tilde{W}$ is a Wiener process and $\tilde{N}$ is a Poisson random measure over the same $(\tilde{X},\tilde{\underline{H}})$, then $\tilde{W}$ and $\tilde{N}$ are necessarily independent of each other (but both may depend on X).

REPRESENTATIONS FOR SEMIMARTINGALES

The following is the main result for m-dimensional additive quasi-left-continuous semimartingales over $(X,\underline{H})$. The setup and notations are exactly as in Theorem (4). The condition on (A_t) will be removed later.

(6) THEOREM. Suppose $A_t = t$ identically. Then, there exists a strong Markov extension $(\tilde{X},\tilde{\underline{H}})$ of $(X,\underline{H})$ supporting a Wiener process $\tilde{W}$ on $\mathbb{R}^m$ and a Poisson random measure $\tilde{N}$ on $\mathbb{R}_+ \times \mathbb{R}$ with mean measure ds $\nu(dz)$ such that

$$
\begin{aligned}
(7) \qquad Y_t = &\int_0^t b(X_s)\, ds + \int_0^t c(X_s)\, d\tilde{W}_s \\
&+ \int_0^t \int_{\mathbb{R}} k(X_{s-},z)\, I_{\{|k(X_{s-},z)|\le 1\}}\, [\tilde{N}(ds,dz) - ds\ \nu(dz)] \\
&+ \int_0^t \int_{\mathbb{R}} k(X_{s-},z)\, I_{\{|k(X_{s-},z)|>1\}}\, \tilde{N}(ds,dz)
\end{aligned}
$$

almost surely under every $\tilde{P}_x$. The decomposition (2) of Y as $Y^b + Y^c + Y^d + Y^e$ does not depend on x, is the same on $(X,\underline{H})$ and on $(\tilde{X},\tilde{\underline{H}})$, and is given by the right-hand-side of (7) in that order. Moreover, we can take $(\tilde{X},\tilde{\underline{H}}) = (X,\underline{H})$ provided that, for each x, the rank of the matrix c(x) be m and the measure $B \to \int \nu(dz)\ 1_B \circ k(x,z)$ be infinite and diffuse.

The proof is quite long and technical, the main difficulty stemming from the construction of the extension $(\tilde{X},\tilde{\underline{H}})$ supporting $\tilde{W}$ and $\tilde{N}$. We refer to [2] for the proof. The following are the ideas behind it.

For simplicity, suppose Y takes values in $\mathbb{R}$. In the decomposition (2), we have $Y^b = B$, which is the same as the first term on the right-side of (7) in view of (5) and our hypothesis regarding A. Concerning the term Y^c in (2), note that, if c never vanishes, then $d\tilde{W}_t = c(X_t)^{-1}dY_t^c$ defines a Wiener process $\tilde{W}$ over $(X,\underline{H})$ and we have $dY_t^c = c(X_t)d\tilde{W}_t$. However, if $c(x) = 0$ for some x, it is impossible to determine $d\tilde{W}_t$ from dY_t^c when $c(X_t) = 0$. Then, the idea is to put

$$d\tilde{W}_t = c(X_t)^{-1}\, I_{\{c(X_t)\neq 0\}}\, dY_t^c + I_{\{c(X_t)=0\}}\, dW_t$$

where W is an auxiliary Wiener process independent of X, constructed on the auxiliary space $(\Omega',\underline{H}',P')$.

For the construction of the Poisson random measure $\tilde{N}$ we work with the random measure Γ defined by (3). To see what must be done, suppose $\tilde{N}$ is already constructed, and note that (7) implies that Γ is the image of $\tilde{N}$ under the mapping $(x,z) \to (s,k(X_{s-},z))$ from $\mathbb{R}_+ \times \mathbb{R}$ into $\mathbb{R}_+ \times \mathbb{R}^m$. Fix $t \geq 0$, $\omega \in \Omega$, $\omega' \in \Omega'$, and set $x = X_{t-}(\omega)$, $y = Y_t(\omega)-Y_{t-}(\omega)$. There are three cases. a) If $y \neq 0$ and $k(x,z) = y$ for exactly one z, say $z = \hat{k}(x,y)$, then $\tilde{N}(\omega,\omega';\{t\}\times B) = 1_B\circ\hat{k}(x,y)$, which does not depend on ω' at all. b) If $y \neq 0$ but $k(x,z) = y$ for all z in some non-singleton set D_{xy}, then we know that $\tilde{N}(\omega,\omega';\{t\}\times D_{xy}) = 1$. Given this information, the actual location of the corresponding atom is some point $z_0(\omega')$ where $w \to z_0(w)$ has the distribution $\nu(dz)/\nu(D_{xy})$ on D_{xy}. Thus,

$$\tilde{N}(\omega,\omega';\{t\}\times B) = 1_B\circ\hat{k}(x,y,U(\omega')), \qquad B \in \underline{\underline{R}},$$

where U is defined on Ω' so that it has the uniform distribution on

(0,1] and where $\hat{k}$ is selected so that $\hat{k}(x,y,U)$ has the distribution $\nu(dz)/\nu(D_{xy})$ on the set D_{xy}. c) If $y = 0$ but $D_x = \{z\colon k(x,z) = 0\}$ is not empty, then we have no information on $\tilde{N}(\omega,\omega';\{t\}\times B)$, and we set

$$\tilde{N}(\omega,\omega';\{t\}\times B) = N(\omega';\{t\}\times(B\cap D_x)), \qquad B \in \underline{\underline{R}},$$

where N is an auxiliary Poisson random measure on the auxiliary space $(\Omega',\underline{\underline{H}}',P')$.

<u>General case</u>. For the representation (7) to hold, it is necessary that $A_t = t$ identically. However, the general case is reduced to this particular case by a random time change as follows. Let

$$\hat{A}_u = \inf\{t\colon\ A_t > u\}, \tag{8}$$

$$\hat{X}_u = X_{\hat{A}_u},\quad \hat{\underline{\underline{F}}}_u = \underline{\underline{F}}_{\hat{A}_u},\quad \hat{\theta}_u = \theta_{\hat{A}_u},\quad \hat{H}_u = \underline{\underline{H}}_{\hat{A}_u},\quad \hat{Y}_u = Y_{\hat{A}_u}. \tag{9}$$

Since A is a strictly increasing and continuous additive functional of X with $\lim_t A_t = \infty$, $\hat{X} = (\Omega,\underline{\underline{F}}_\infty,\hat{\underline{\underline{F}}}_u,\hat{\theta}_u,\hat{X}_u,P_x)$ is a Markov process with the same properties as X, $\hat{\underline{\underline{H}}} = (\hat{\underline{\underline{H}}}_u)$ is a strong Markov filtration for $\hat{X}$, and $\hat{Y}$ is an additive quasi-left-continuous semimartingale over $(\hat{X},\hat{\underline{\underline{H}}})$. Moreover, the local characteristics $\hat{B}$, $\hat{C}$, $\hat{G}$ of Y over $(\hat{X},\hat{\underline{\underline{H}}})$ satisfy (5); for example,

$$\hat{B}_u = B_{\hat{A}_u} = \int_0^{\hat{A}_u} b(X_s)\, dA_s = \int_0^u b(X_{\hat{A}_s})\, ds = \int_0^u b(\hat{X}_s)\, ds.$$

Thus, Theorem (6) apply to the process $\hat{Y}$ over $(\hat{X},\hat{\underline{\underline{H}}})$: there is a strong Markov extension $(\tilde{X},\tilde{\underline{\underline{H}}})$ of $(\hat{X},\hat{\underline{\underline{H}}})$ supporting a Wiener process $\tilde{W}$ and a Poisson random measure $\tilde{N}$ such that (7) holds for $\hat{Y}$ and $\hat{X}$. Of course, Y is related to $\hat{Y}$ by $Y_t = \hat{Y}_{A_t}$.

REPRESENTATION FOR MARKOV PROCESSES

We now describe the implications of the results mentioned in the preceding section for the Markov process (X_t) itself, supposing that the state space is $E = \mathbb{R}^m$ and that X is a quasi-left-continuous semimartingale with respect to every P_x. Then, we call X a <u>semimartingale Hunt process</u>. For such a process X, $(Y_t) = (X_t - X_0)$ is an additive quasi-left-continuous semimartingale over $(X,\underline{\underline{H}})$ where we can now take $\underline{\underline{H}}^o_t = \underline{\underline{F}}^o_t$. Let A, b, c, k be related to $Y = X - X_0$ through Theorem (4). If $A_t = t$ identically, we call X an <u>Ito process</u>.

In view of the remarks at the end of the preceding section, every semimartingale Hunt process X is obtained by a random time change from an Ito process $\hat{X}$; in other words, $X_t = \hat{X}_{A_t}$ where $A_t = \inf\{u: \hat{A}_u > t\}$ for a strictly increasing continuous additive functional $\hat{A}$ of the Ito process $\hat{X}$. Moreover, since $dt \ll dA_t$ (see Theorem (4)), the additive functional $\hat{A}$ is of a simple type:

(10) $$\hat{A}_u = \int_0^u a(\hat{X}_s)\, ds$$

for some positive bounded Borel function a. These remarks allow us to concentrate on Ito processes.

The following is obtained from Theorem (6) by taking $Y = X - X_0$. In particular, it shows that these are precisely the class of processes that ITO [11] introduced with some extra conditions on b, c, k to ensure existence and uniqueness.

(11) THEOREM. Let X have state space $E = \mathbb{R}^m$, and set $\underline{\underline{H}}^o_t = \underline{\underline{F}}^o_t$. Then, X is an Ito process if and only if there exists a strong Markov extension $(\tilde{X},\tilde{\underline{\underline{H}}})$ of $(X,\underline{\underline{H}})$ supporting a Wiener process $\tilde{W}$ on $\mathbb{R}^m$ and a Poisson random measure $\tilde{N}$ on $\mathbb{R}_+\times\mathbb{R}$ with mean measure $ds\, \nu(dz)$ such that (X_t) satisfies

$$(12) \qquad X_t = X_0 + \int_0^t b(X_s)\,ds + \int_0^t c(X_s)\,d\tilde{W}_s$$

$$+ \int_0^t \int_{\mathbb{R}} k(X_{s-},z)\, I_{\{|k(X_{s-},z)| \leq 1\}} [\tilde{N}(ds,dz) - ds\, \nu(dz)]$$

$$+ \int_0^t \int_{\mathbb{R}} k(X_{s-},z)\, I_{\{|k(X_{s-},z)| \geq 1\}} \tilde{N}(ds,dz)$$

$\tilde{P}_x$ - almost surely for every $x \in E$ for some Borel functions b, c, and k.

Processes with paths of locally bounded variation. Let X be a Hunt process whose paths are of locally bounded variation. Then X is automatically a semimartingale, and hence, it is obtained by a random time change from an Ito process $\hat{X}$ whose paths are of locally bounded variation. The preceding theorem applies to $\hat{X}$, but now (12) becomes simpler: $c = 0$ and k is such that we can write

$$(13) \qquad \hat{X}_t = \hat{X}_0 + \int_0^t \hat{b}(\hat{X}_s)\,ds + \int_0^t \int_{\mathbb{R}} k(\hat{X}_{s-},z)\, \tilde{N}(ds,dz)$$

by letting $\hat{b}(x) = b(x) - \int \nu(dz)\, k(x,z)\, I_{\{|k(x,z)| \leq 1\}}$.

If X is further continuous, then $k = 0$ in (13). Since the only strong Markov solutions of (13) with $k = 0$ are deterministic, except for the choice of $\hat{X}_0$, the processes $\hat{X}$ and therefore X are basically deterministic. In other words, if X is a continuous Hunt process whose paths are of locally bounded variation, then there exists a deterministic Borel function $(x,t) \to p(x,t)$ such that $P_x\{\omega:\ X_t(\omega) = p(x,t)$ for all $t \geq 0\} = 1$ for every x.

Existence-Uniqueness. The equations like (12) were introduced by ITO [11] with conditions on b, c, k to ensure existence and uniqueness of a solution X (given b, c, k, $\tilde{W}$, and $\tilde{N}$). In a weaker sense, STROOCK [16] and STROOCK and VARADHAN [17] give sufficient conditions on b, c, k to ensure the existence of a Markov process X such that (12) is satisfied for some $\tilde{W}$ and $\tilde{N}$. Theorem (11) is a converse to such results: it shows that for every Ito process X such an equation (12) holds for some b, c, k, $\tilde{W}$, and $\tilde{N}$. Our necessary conditions on b, c, k to be so related to a process X are weaker than the known sufficient conditions.

Examples. Every process with stationary and independent increments is an Ito process where the coefficients b, c, k satisfy $b(x) = b_0$, $c(x) \doteq c_0$, $k(x,z) = k_0(z)$ independent of x, and $\int \nu(dz)(|k_0(z)|^2 \wedge 1) < \infty$. Ito processes with $k = 0$ are called diffusions (or quasi-diffusions), and are studied extensively. Every regular step process on $\mathbb{R}^m$ is an Ito process satisfying (13) with $\hat{b}(x) = 0$ and $\int \nu(dz)\, 1_D \circ k(x,z) < \infty$ where $D = \mathbb{R}^m \setminus \{0\}$. Absolute value of the standard Brownian motion on $\mathbb{R}$ is a semimartingale Hunt process, but is not an Ito process; in its representation we have $b(x) = 1_{\{0\}}(x)$, $c(x) = a(x) = 1_{(0,\infty)}(x)$.

Finally, we should mention that there are Hunt processes that are not semimartingales, for instance, $X_t = |W_t|^{\frac{1}{2}}$ is a continuous Hunt process which is not a semimartingale (see YOR [18] for this). Our representation theorems fail to hold for such processes. However, sometimes, it is possible to find deterministic functions $f_1,\ldots,f_m$ such that each $f_i(X)$ is a semimartingale and $\{f_1,\ldots,f_m\}$ separates the points of the state space (if $E = \mathbb{R}$ and X is continuous, then FELLER has shown that there exists one such function f). Then, Theorem (6) applies to $Y_t = (f_1\circ X_t - f_1\circ X_0,\ldots,f_m\circ X_t - f_m\circ X_0)$, and the resulting representation may be used for studying X. Deterministic functions f such that f(X) is a semimartingale were characterized in [3]. Whether there exists a finite (or at least countable) collection of such functions that further separate the points of E is a harder question. KNIGHT [13] gives some answers to this question.

REFERENCES

[1] R.M. BLUMENTHAL and R.K. GETOOR. Markov Processes and Potential Theory. Academic Press, New York, 1968.

[2] E. ÇINLAR and J. JACOD. Representation of semimartingale Markov processes in terms of Wiener processes and Poisson random measures. 1980, pp. 92, to appear.

[3] E. ÇINLAR, J. JACOD, P. PROTTER, and M.J. SHARPE. Semimartingales and Markov processes. Z. Wahrscheinlichkeitstheorie verw. Gebiete.

[4] E.B. DYNKIN. Markov Processes. Academic Press, New York, 1965.

[5] N. EL KAROUI and J.-P. LEPELTIER. Représentation des processus ponctuels multivariés à l'aide d'un processus de Poisson. Z. Wahrscheinlichkeitstheorie verw. Gebiete, 39 (1977), 111-133.

[6] W. FELLER. The general diffusion operator and positivity preserving semigroups in one dimension. Ann. Math. 60 (1954), 417-436.

[7] W. FELLER. On second order differential operators. Ann. Math. 61 (1955), 90-105.

[8] W. FELLER. Generalized second order differential operators and their lateral conditions. Illinois J. Math. 1 (1957), 495-504.

[9] B. GRIGELIONIS. On the representation of integer-valued random measures by means of stochastic integrals with respect to the Poisson measure. Lit. Math. J. 11 (1971), 93-108.

[10] K. ITO. On stochastic processes (I) (Infinitely divisible laws of probability). Japan J. Math. 18 (1942), 261-301.

[11] K. ITO. On Stochastic Differential Equations. Mem. Amer. Math. Soc. 4 (1951).

[12] J. JACOD. Calcul Stochastique et Problèmes de Martingales. Lecture Notes in Math. 714, Springer-Verlag, Berlin, 1979.

[13] F.B. KNIGHT. An infinitesimal decomposition for a class of Markov processes. Ann. Math. Statist. 41 (1970), 1510-1529.

[14] A.V. SKOROKHOD. On homogeneous continuous Markov processes that are martingales. Theo. Prob. Appl. 8 (1963), 355-365.

[15] A.V. SKOROKHOD. On the local structure of continuous Markov processes. Theo. Prob. Appl. 11 (1966), 366-372.

[16] D.W. STROOCK. Diffusion processes associated with Lévy generators. Z. Wahrscheinlichkeitstheorie verw. Gebiete 32 (1975), 209-244.

[17] D.W. STROOCK and S.R.S. VARADHAN. Multidimensional Diffusion Processes. Springer-Verlag, Berlin, 1979.

[18] M. YOR. Un exemple de processus qui n'est pas une semimartingale. Temps Locaux, pp. 219-222. Astérisque, No. 52-53 (1978).

E. Cinlar
Northwestern University
Evanston Illinois 60201.
U.S.A.

J. Jacod
Université de Rennes
Beaulier, 35042 - Rennes- Cedex
FRANCE

THE EXPECTED VALUE OF PERFECT INFORMATION

IN THE OPTIMAL EVOLUTION OF STOCHASTIC SYSTEMS

M.A.H. Dempster
System and Decision Sciences, IIASA,
A-2361 Laxenburg, Austria

(On leave from Balliol College, University of Oxford, Oxford OX1 3BJ, England)

CONTENTS

1. INTRODUCTION

This paper uses abstract optimization theory to characterize and analyze the stochastic process describing the current *marginal expected value of perfect information* in a class of discrete time dynamic stochastic optimization problems which includes the familiar optimal control problem with an infinite planning horizon. Using abstract Lagrange multiplier techniques on the usual nonanticipativity constraints treated *explicitly* in terms of adaptation of the decision sequence, it is shown that the marginal expected value of perfect information is a supermartingale. For a given problem, the statistics of this process are of fundamental practical importance in deciding the necessity for continuing to take account of the stochastic variation in the evolution of the sequence of optimal decisions.

Let $\underset{\sim}{x} = \{\underset{\sim}{x}_t\}_{t=1}^{\tau}$ be a sequence of *decisions* in $\mathbb{R}_n$ and let $\underset{\sim}{\xi} = \{\underset{\sim}{\xi}_t\}_{t=1}^{\tau}$ be a discrete time stochastic process in (Ξ,Σ,μ) of subsequent *observations*. A *policy (decision rule or recourse function)* is a measurable map $\underset{\sim}{x}:\xi \to x(\xi)$. Consider the problem

(RP) $$\inf_{\underset{\sim}{x}} E[\Sigma_{t=1}^{\tau} \underset{\sim}{f}_t(\underset{\sim}{x})]$$

s.t. $\underset{\sim}{x}_t \in P_t \subset \mathbb{R}^{n_t}$ $\quad \underset{\sim}{g}_t(\underset{\sim}{x}) \in Q_t \subset \mathbb{R}^{m_t}$ $\quad$ a.s. $t = 1,\ldots,\tau$

(where $n_t \leq n_\tau = n$, $m_t \leq m_\tau = m$) and the *nonanticipative* condition that the current decision $\underset{\sim}{x}_t$ depends only on the sequence of observations $\xi_1,\xi_2,\ldots,\xi_{t-1}$, and realised decisions $x_1,x_2,\ldots,x_{t-1}$ (and thus $\underset{\sim}{\xi}$) to date. Here $f_t\colon \Xi \times X_{t=1}^{\tau} \mathbb{R}^{n_t} \to \mathbb{R}$ is assumed measurable in its first argument and Borel measurable in its second and $\underset{\sim}{f}_t(\underset{\sim}{x}) := f_t(\cdot,x(\cdot))$, and similarly for $\underset{\sim}{g}_t(\underset{\sim}{x})$. (A full set of technical assumptions will be introduced in §§2 and 3.)

The problem (RP)--termed the *dynamic recourse problem*--has a number of important applications in the mathematical sciences (*cf.* Dempster, 1980). Special cases include stochastic dynamic linear or quadratic programming formulations of *energy-economic planning* models, Birge (1980), Louveaux and Smeers (1980,1981); *inventory control* models, see e.g. Veinott (1966); *Markov decision processes* with random transition matrices for *manpower planning*, Grinold (1976,1980) and the classical *discrete time optimal control* model. To see the last assertion in more detail, make the following substitutions in (RP):

$$\underset{\sim}{x} := (\underset{\sim}{z},\underset{\sim}{u}) \qquad m_t :\equiv m \qquad n_t :\equiv m+n$$

$$Q_t :\equiv \{0\} \qquad P_1 := \{z_1\} \times \mathbb{R}^n \qquad P_t := \mathbb{R}^{m+n} \quad t=2,\ldots,\tau$$

$$\underset{\sim}{f}_t(\underset{\sim}{x}) := \underset{\sim}{f}_t(\underset{\sim}{x}_t) \qquad \underset{\sim}{g}_\tau(\underset{\sim}{x}) :\equiv 0$$

$$\underset{\sim}{g}_t(\underset{\sim}{x}) := \underset{\sim}{z}_{t+1} - h(\underset{\sim}{z}_t,\underset{\sim}{u}_t) - \underset{\sim}{\xi}_t \qquad t=1,\ldots,\tau-1 \quad .$$

Then (RP) reduces to the familiar *control problem*

(C) $$\sup_{\underset{\sim}{u}} E[\Sigma_{t=1}^{\tau} \underset{\sim}{f}_t(\underset{\sim}{z}_t,\underset{\sim}{u}_t)]$$

s.t. $$\underset{\sim}{z}_{t+1} = h(\underset{\sim}{z}_t,\underset{\sim}{u}_t) + \underset{\sim}{\xi}_t \qquad \text{a.s.} \qquad t=1,\ldots,\tau-1$$

$$\underset{\sim}{z}_1 = z_1 \qquad \text{a.s.} \quad .$$

Control and state space constraints are easily added to (C) by suitable definition of P_t, $t=1,\ldots,\tau$.

Characterization of the (optimal) solutions to the general problem (RP) for finite τ has been treated by Rockafellar and Wets(1976a,b, 1978) for the convex case under a Slater *regularity condition (constraint qualification)* using the duality theory of convex conjugate functions. More recently (1981), they have given a similar treatment of the convex Bolza problem--a special case of (C)--for finite τ . Hiriart-Urruty (1978,1981) has considered a more general class of nonlinear special cases of (RP) for finite τ . He applied a theory characterizing in terms of generalized gradients the minimum of an integral functional involving a measurable locally Lipschitz integrand subject to a measurable closed valued multifunction constraint. The version of (RP) he treated has some nondifferentiable and some differentiable constraints and a correspondingly mixed Slater/Mangasarian-Fromowitz constraint qualification is enforced.

This paper discusses the use of recently developed abstract mathematical programming theory (Dempster,1976; Zowe and Kurcyusz,1979; Brokate,1980) to extend these results to the *infinite horizon* (τ infinite) problem for the general nonlinear case under appropriate regularity conditions involving problem functions in L_∞ . The *support representation* problem (Dempster,1976) for (RP) is addressed to obtain the appropriate stochastic maximum principle and nonanticipative supermartingale representation of the L_1 multiplier process corresponding to the nonanticipative constraints. The general results were announced in Dempster (1980); details and proofs will appear elsewhere. In this paper, emphasis is placed on precise problem formulation, results and interpretation (§§2 and 3) and attention is focused on the supermartingale multiplier process corresponding to the nonanticipative constraints (§4). Technical limitations to extending the analysis of this marginal expected value of perfect information process to continuous time systems --involving say diffusion or jump dynamics--are discussed briefly in §5 with a view to their possible relaxation in future work using recent theories of pathwise integration of stochastic differential equations (Sussman,1978; Marcus,1981).

2. THE DYNAMIC RECOURSE PROBLEM AS AN ABSTRACT OPTIMIZATION PROBLEM

The purpose of this section is to set up the problem (RP) as an abstract mathematical programme of the form

$$\text{(P)} \qquad \sup_{x\in P} f(x) \qquad \text{s.t.} \qquad g(x) \in Q \qquad ,$$

where P and Q are sets in appropriate linear topological spaces U and V, $f:U \to \mathbb{R}$ and $g:U \to V$.

A natural assumption to make in the applications cited in §1 is that feasible policies should be *essentially bounded*--i.e. in L_∞ -- on (Ξ,Σ,μ), *cf.* Rockafellar and Wets (1976,1978,1981), although a similar treatment of policies in $L_p (1 \leq p < \infty)$ is also possible, *cf.* Eisner and Olsen (1975,1980), Hiriart-Urruty (1978,1981). Hence assume Σ completed with respect to μ and define

$$(2.1) \qquad L_\infty^n := L_\infty[(\Xi,\Sigma,\mu);\mathbb{R}^n]$$

and

$$x := \underset{\sim}{x} = (\underset{\sim}{x}_1,\ldots,\underset{\sim}{x}_\tau) \qquad U := X_{t=1} L_\infty^{n_t}$$

$$P_t^\Xi := \{\phi \varepsilon L_\infty^{n_t} \mid \phi:\Xi \to P_t\} \qquad P := X_{t=1}^\tau P_t^\Xi$$

$$Q_t^\Xi := \{\Psi \varepsilon L_\infty^{m_t} \mid \Psi:\Xi \to Q_t\} \qquad Q := X_{t=1}^\tau Q_t^\Xi \quad ,$$

equipping U with the usual equivalent of the product topology defined in terms of the norm $\|u\| = \sup_t \|u_t\|_\infty$. (We shall be party to the usual abuse of terminology by referring to the equivalence class elements of Banach function spaces as functions and subject to the standard analyst/probabilist's schizophrenia by denoting the elements of, for example, L_∞ as both u and $\underset{\sim}{u}$ depending on whether the analytic or probabilistic interpretation is to the fore.) The abstract objective function of (P) is obviously

$$(2.2) \qquad f(x) := E[\Sigma_{t=1}^\tau \underset{\sim}{f}_t(\underset{\sim}{x})] \quad ,$$

but a little more analysis will be necessary in order to define the abstract constraint function g and its range (image) space V.

The problem lies with the *explicit* characterization of nonanticipation. Its solution was first proposed in the context of stochastic *linear* programming by Eisner and Olsen (1975,1980). Let $\{\Sigma_t\}$ be the usual increasing tower of σ-fields,

$$\Sigma_t \subset \Sigma_\tau := \Sigma \quad ,$$

generated by the process $\underset{\sim}{\xi}$. Every feasible policy $\underset{\sim}{x}$ is assumed to be $\underset{\sim}{\xi}$ *adapted*, i.e.

(2.3) $\quad \underset{\sim}{x}_t = E\{\underset{\sim}{x}_t \mid \Sigma_{t-1}\}$ $\quad$ a.s. $\quad$ $t=1,\ldots,\tau$.

After canonically embedding $L_\infty^{n_t}$ in L_∞^n in the obvious way, we may define the closed projections

$$\Pi_t : L_\infty^n \to L_\infty^n \ , \ z \mapsto \Pi_t z := E\{\underset{\sim}{z} \mid \Sigma_t\}$$

for $t=1,\ldots,\tau$. Then $\underset{\sim}{x}$ is adapted to $\underset{\sim}{\xi}$ if, and only if,

(2.3') $\quad (I - \Pi_{t-1}) x_t = 0$ $\quad$ $t=1,\ldots,\tau$.

Here (when we make the usual assumption that x_1 is deterministic) $\Pi_0 := \Pi_{K_1}$, where K_1 denotes the linear span of the constants in $L_\infty^{n_1}$ embedded in L_∞^n .

We may now define the constraint function of (P) as

(2.4) $\quad g: U \mapsto V := \text{X}_{t=1}^{\tau} L_\infty^{m_t} \times (L_\infty^n)^\tau$

such that

$$x \to g(x) := (\underset{\sim}{g}_1(x), \underset{\sim}{g}_2(x), \ldots \ ; \ (I-\Pi_0)x_1, \ (I-\Pi_1)x_2, \ldots \ ) \ ,$$

with V equipped with the normed-defined equivalent of the product topology (as with U). All these considerations apply equally well to finite or infinite τ . In the sequel we shall consider only the case of infinite τ ; necessary changes for the simpler case of finite τ are easily supplied by the reader. Further, define $\underset{\sim}{\xi}^t, \underset{\sim}{x}^t$ to be the *history*, i.e. $\underset{\sim}{\xi}_1, \underset{\sim}{\xi}_2, \ldots, \underset{\sim}{\xi}_t; \ \underset{\sim}{x}_1, \underset{\sim}{x}_2, \ldots, \underset{\sim}{x}_t$, of the observation, respectively policy, process to time t . We shall specialize in what follows to the case--relevant to all the practical examples of §1--of *triangular* (RP), for which current constraints depend only on observation and decision histories *to date*, i.e.

(2.5) $\quad g_t(\underset{\sim}{\xi}^t, \underset{\sim}{x}^t) \in Q_t$ $\quad$ a.s. $\quad$ $t=1,2,\ldots$.

(Thus the measurability properties for the g_t cited in §1 may be restricted to Σ_t rather than Σ .)

This completes the basic set of assumptions needed to specify and analyze the dynamic recourse problem (RP) over an infinite horizon. Further (illustrative) technical assumptions will be introduced as required to complete the analysis in §3.

3. CHARACTERIZATION OF OPTIMAL SOLUTIONS

The characterization of optimal solutions for the abstract mathematical programming problem (P) necessitates consideration (perhaps only implicitly) of its *Langrangian function*

$$(3.1)\qquad \phi(x,y') := f(x) + y'g(x) \quad ,$$

for $x \in U$ and *multiplier* vectors $y' \in V'$, the *dual* space of V (consisting of all linear functionals on V continuous in the given topology for V).

We shall thus here need the following characterization of L_∞^*, due essentially to Yosida and Hewitt (1952) (see also Dubovitskii and Milyutin, 1965, Valadier, 1974, and Dempster, 1976). A finitely additive row n-vector valued measure $\pi' : \Sigma \to \mathbb{R}^{n'}$ on (Ξ,Σ) is *purely finitely additive (p.f.a.)* if, and only if, for all countably additive real valued measures ν on (Ξ,Σ) and for all $\varepsilon>0$ there exists $A_\varepsilon \in \Sigma$ such that $|\nu(A_\varepsilon)| < \varepsilon$ and $\pi'(A_\varepsilon) = 0' \in \mathbb{R}^{n'}$, i.e. a p.f.a. measure is carried by sets assigned arbitrarily small measure by any countably additive measure. (Prime is used to denote a dual element; in the finite dimensional case this is consistent with vector transposition.) Denote by $L_\infty^{n^*}$ the (Banach) dual space of L_∞^n (as defined in (2.1)), by $L_1^{n'}$ the space of (coordinatewise) absolutely integrable row n-vector valued functions on (Ξ,Σ) and by $P^{n'}$ the space of purely finitely additive row n-vector valued measures on (Ξ,Σ).

Proposition: 3.1. Given the measure space (Ξ,Σ,μ), if Σ is complete with respect to the σ-finite measure μ, then

$$L_\infty^{n^*} \cong L_1^{n'} \oplus P^{n'} \qquad \blacksquare$$

Here $\cong$ denotes isometric isomorphism and the action of $y' \in L_\infty^{n^*}$ on an n-vector valued function $x \in L_\infty^n$ is given by

$$(3.2)\qquad y'x := \int_\Xi y'(\xi)x(\xi)\mu(d\xi) + \int_\Xi \pi'(d\xi)x(\xi)$$

$$:= y_1'x + y_2'x \quad .$$

The first integral in (3.2) is simply an abstract Lebesque integral; the second requires the analogous integration theory developed for *finitely* additive measures by Dunford and Schwartz (1956). In fact, Valadier (1974) extended the result of Proposition 3.1. to the σ-finite case from the finite case established by Yosida and Hewitt (1952), while Dubovitskii and Milyutin (1967) independently gave a complete treatment of L_∞^* in terms of *singular functionals* (y_2' of (3.2)) without reference to their integral representations. A finer characterization of L_∞^* in terms of natural subspace of $P^{n'}$ appears in Dempster (1976).

We are now in a position to make precise sense of the Langrangian function (3.1) for (P). According to (2.4) we are interested in representation of the dual space, $X_{t=1}^{\infty} L_\infty^{m_t *} \times (L_\infty^n)^{\mathbb{N}}$, of V $(=X_{t=1}^{\infty} L_\infty^{m_t} \times (L_\infty^n)^{\mathbb{N}})$, where $\mathbb{N}$ denotes the natural numbers. A straightforward application of Proposition 3.1 yields ϕ as given by

(3.3) $\quad \phi(x,y')$

$$= \Sigma_{t=1}^{\infty} \{E\underset{\sim}{f}_t(\underset{\sim}{x}) + E\underset{\sim}{y}_t'\underset{\sim}{g}_t(\underset{\sim}{x}) + E\underset{\sim}{\rho}_t'(I-\Pi_{t-1})\underset{\sim}{x}_t$$

$$+ \int_{\Xi} \pi_t'(d\xi) g_t(\xi,x(\xi))$$

$$+ \int_{\Xi} \Psi_t'(d\xi)\ [x_t(\xi) - E\{x_t|\Sigma_{t-1}\}(\xi)]$$

$$+ \int_{\Xi\times\mathbb{N}} \chi_\infty'(d\xi,dt)\ (g_t(\xi,x(\xi)),\ \underset{\sim}{x}_t(\xi) - E\{\underset{\sim}{x}_t|\Sigma_{t-1}\}(\xi))$$

using the fact (Yosida and Hewitt, 1952) that all p.f.a. measures on $\mathbb{N}$ (with counting measure # taken as ground measure) are carried by neighbourhoods of ∞. In (3.3)

$$\underset{\sim}{y}_t' \in L_\infty^{n_t'},\quad \underset{\sim}{\rho}_t' \in L_\infty^{m_t'},\quad \pi_t' \in P^{n_t'},\quad \Psi_t' \in P^{n'},$$

$\chi_\infty' \in P[(\Xi\times\mathbb{N},\ \varepsilon\times\gamma(\mathbb{N}),\ \mu\times\#)\ ;\ \mathbb{R}^{m+n'}]$, the space of row $(m+n)$-vector valued measures on the product σ-field shown, and in the corresponding integral g_t has been canonically embedded in $\mathbb{R}^m$.

Next we characterize an optimum x_o of (P) in terms of a suitable concept of derivatives of the Lagrangian ϕ given by (3.3). Rather than use minimal concepts and introduce highly technical conditions on (P), we shall by way of illustration use Fréchet derivatives and give regularity conditions only for (RP) sufficient to ensure the truth of the following *Kuhn-Tucker Theorem* for (P), *cf.* Zowe and Kurcyusz (1979).

Suffice it to say here that versions of Proposition 3.2 below are available involving both generalized derivatives (*cf.* Hiriart-Urruty, 1981) and (one-sided) Gateaux directional derivatives (Dempster, 1976) under *minimal* regularity conditions (*cf.* Dempster, 1976; Brokate, 1980) for (P) posed in locally convex Hausdorff topological vector spaces. We shall need the concept of the *dual cone* $Q' \subset V'$ of a set $Q \subset V$ as

$$Q' := \{y' \in V' : y'z \geqq 0,\ \forall z \in Q\} \quad , \tag{3.4}$$

and similarly for sets in U.

Proposition: 3.2. Let U and V be Banach spaces and the problem functions f and g of (P) be Fréchet differentiable with derivatives ∇f and ∇g respectively. Then under suitable regularity conditions on (P), x_0 an optimum for (P) implies that there exists $y_0' \in Q'$ such that

$$\omega_0' := \nabla_x \phi(x_0, y_0') = \nabla f(x_0) + y_0' \nabla g(x_0) \in -P'$$

$$g(x_0) = \nabla_{y'} \phi(x_0, y_0') \in Q \tag{3.5}$$

$$\omega_0' u \leqq \omega_0' x_0 \qquad \forall u \in P$$

$$y_0' g(x_0) \leqq y_0' v \qquad \forall v \in Q \quad . \qquad \blacksquare$$

Conditions (3.5) are termed *Kuhn-Tucker (necessary) conditions* for an optimum of (P).

If $0 \in P \subset U$ and $0 \in Q \subset V$, the last two imply the *complementary slackness* conditions

$$\omega_0' x_0 = 0 \qquad y_0' g(x_0) = 0 \quad . \tag{3.6}$$

Applying Proposition 3.2 under suitable regularity assumptions on (RP) yields a (necessary) characterization of its optimal policies in terms of the *abstract* Lagrangian of (P) given by (3.3). However, inspection of (3.3) raises the question of conditions under which this characterization remains valid if the awkward terms involving integrals with respect to purely finitely additive measures are dropped. This is a special case of the general *support representation problem* (Dempster, 1976) which has appeared in the control (Dubovitskii and Milyutin, 1965), economics (e.g. Prescott and Lucas, 1972) and optimization (Rockafellar and Wets, 1976a,b, 1978) literature.

To solve the support representation problem for (RP) we must give conditions on the problem sufficient to make both the *stochastic* p.f.a. measures π_t' and ψ_t', t=1,2,..., and the *intertemporal* p.f.a. measure χ_∞' of (3.3) vanish. The following conditions are a distillation of the literature cited above. Some terms and definitions will be needed. A policy history $\underset{\sim}{x}^t$ is termed *feasible* if it satisfies the constraint structure of (RP) to time t, i.e. if we have for its components

$$\underset{\sim}{x}_s \in P_s \qquad \underset{\sim}{g}_s(\underset{\sim}{x}^s) \in Q_s \quad \text{a.s.} \qquad s = 1,\dots,t \ .$$

Define

$$X_t := \{x_t \in \mathbb{R}^{n_t} : x_t = x_t(\xi) \ , \xi \in \Xi \ , \underset{\sim}{x}^t \text{ feasible}\}$$

and

$$C_t := \{x_t \in \mathbb{R}^{n_t} : x_t = x_t(\xi) \ , \xi \in \Xi, \ \underset{\sim}{x} \text{ feasible}\} \ .$$

Clearly C_t essentially lies in X_t, but the analytical (computational) intractability of (RP) arises from the fact that this inclusion is in general essentially strict. The *controllability* (or *relatively complete recourse*) condition

$$(3.7) \qquad \mu|_{\Sigma_t}(X_t \Delta C_t) = 0 \qquad t = 1,2,\dots \quad ,$$

ensures almost sure decision recourse at all times from any realization of the observation (and decision) process to date and forces the optimal stochastic p.f.a. measure π_t' of Proposition 3.2 to vanish. Rockafellar and Wets (1976a) obtain more technical sufficient conditions. They show by example that, without the *explicit* introduction into the problem of the constraints which bind at an optimum induced on C_t by later stages, the support representation problem in terms of L_1 multipliers for (RP) is *insoluble*. Since a nonanticipative constraint (2.3) cannot lead to infeasibilities of subsequent nonanticipative constraints, we can conclude immediately that the stochastic p.f.a. measures ψ_t', t=1,2,... always vanish at the optimum. To ensure that the optimal intertemporal p.f.a. measure χ_∞' vanishes, it suffices to assume that $0 \in P_t$, t=1,2,... and the *finite horizon approximation* condition:

(3.8) for some $\tau (\geq 1)$, $\underset{\sim}{x}$ feasible implies $(\underset{\sim}{x}^t,0,0,\dots)$ feasible for all $t \geq \tau$.

Examples of nontrivial p.f.a. measures (in the absence of this condition) are known (see Prescott and Lucas,1972).

Next we must state suitable regularity conditions on (RP) sufficient to ensure the conclusions of Proposition 3.2. in terms of L_1 multipliers. Hence we assume (by way of illustration) that $P_t \subset \mathbb{R}^{n_t}$ and $Q_t \subset \mathbb{R}^{m_t}$ are closed convex cones, $0 \in Q_t$ and the problem functions f_t, g_t are differentiable with respect to their policy components with gradients $\nabla f_t, \nabla g_t$, $t=1,2,\ldots$. Given a set C (in a linear topological space) and a point $x \in C$ define the *inner approximation cone*

$$I(x;C) := \cap_{\mathcal{N}} \cup_{\lambda \geqq 0} \lambda(C-x) \quad ,$$

where the intersection is taken over neighbourhoods $\mathcal{N}$ of x. We shall assume that for an optimal policy $\underset{\sim}{x}_0$

$$(3.9) \qquad \underset{\sim}{g}_t(\underset{\sim}{x}_0^t) + \nabla \underset{\sim}{g}_t(\underset{\sim}{x}_0^t)\, I(\underset{\sim}{x}^t, X_{\tau=1}^{t} P_\tau) - Q_t = \mathbb{R}^{m_t}.$$

$$\text{a.s.} \quad t=1,2,\ldots \quad .$$

Since the (linear) projections $I - \Pi_t$ of (2.3') are closed, (3.9) is sufficient to ensure that the abstract problem function g of (2.4) satisfies

$$(3.10) \qquad g(x_0) + \nabla g(x_0)\, I(x_0;P) - Q = V$$

in terms of its Fréchet derivative at an optimum x_0 of (P) for the closed convex cone $Q \subset V$, *cf.* Dempster (1976), Zowe and Kurcyusz (1979).

Finally we are in a position to apply Proposition 3.2 to obtain a *local* version of the Kuhn-Tucker necessary conditions (3.5) for an optimal policy $\underset{\sim}{x}_0$ of (RP), in terms of a *stochastic maximum principle* involving the L_1 multiplier processes $\underset{\sim}{y}'$ and $\underset{\sim}{\rho}'$, of the form

$$y_t' \in Q_t' \qquad \text{a.s.}$$

$$\Sigma_{s=1}^{\infty} \nabla_{x_t} \underset{\sim}{f}_s(\underset{\sim}{x}) + \Sigma_{s=t}^{\infty} \underset{\sim}{y}_s' \nabla_{x_t} \underset{\sim}{g}_s(\underset{\sim}{x}^s) \in - P_t' \qquad \text{a.s.}$$

$$(3.11) \qquad \underset{\sim}{x}_t \in P_t \quad \underset{\sim}{g}_t(\underset{\sim}{x}^t) \in Q_t \quad \underset{\sim}{x}_t = E\{\underset{\sim}{x}_t | \Sigma_{t-1}\} \qquad \text{a.s.}$$

$$E[\, \Sigma_{s=1}^{\infty} \nabla_{x_t} \underset{\sim}{f}_s(\underset{\sim}{x}) + \Sigma_{\tau=t}^{\infty} \underset{\sim}{y}_s' \nabla_{x_t} \underset{\sim}{g}_s(\underset{\sim}{x}^s)\,]\, \underset{\sim}{x}_t = 0$$

$$E\rho_t'(\underset{\sim}{x}_t - E\{\underset{\sim}{x}_t | \Sigma_{t-1}\}) = 0 \qquad \text{for} \quad t=1,2,\ldots \quad .$$

In the case that all problem functions are concave in the policy variables, conditions (3.11) are also *sufficient*--in general, they are not.

In the next section, we turn to an analysis and interpretation of the (optimal) multiplier process $\underset{\sim}{\rho}'$ corresponding to the nonanticipative constraints (2.3) of (RP).

4. THE MARGINAL EXPECTED VALUE OF PERFECT INFORMATION SUPERMARTINGALE

In this section we shall assume that a fixed optimal policy $\underset{\sim}{x}_0$ for (RP) is specified. (Various further assumptions may be adduced to the problem to guarantee existence and even uniqueness of the optimal policy, but these will not concern us here.) We shall apply modern perturbation theory for the abstract programme (P) of §2, see e.g. Lempio and Maurer (1980), to study the nonanticipative constraint multiplier process $\underset{\sim}{\rho}'$ corresponding to the chosen optimal policy $\underset{\sim}{x}_0$ for (RP). Of interest are perturbations to the nonanticipative constraints (2.3) of the form

$$(4.1) \qquad \underset{\sim}{x}_t = E\{\underset{\sim}{x}_t|\Sigma_{t-1}\} - \underset{\sim}{z}_t \qquad \text{a.s.} \qquad t=1,2,\ldots,$$

where the z_t are arbitrary n-vector valued functions measurable with respect to $\Sigma_{s(t)}$ $(s(t) > t)$ and hence representing information on the *future* of the observation process $\underset{\sim}{\xi}$. More specifically, fix t and an arbitrary n-vector valued function z_t measurable with respect to Σ_s for some fixed $s > t$ and consider perturbations of the tth nonanticipative constraint of the form

$$(4.2) \qquad \underset{\sim}{x}_t = E\{\underset{\sim}{x}_t|\Sigma_{t-1}\} - \alpha\underset{\sim}{z}_t \qquad \text{a.s.}$$

for $\alpha \in [0,\delta]$ $(\delta > 0)$. Denote the optimization problem resulting from the perturbation (4.2) as $(RP[\alpha z_t])$ and its abstract equivalent as $(P[\alpha z_t])$. Define the abstract *perturbation function* $\pi : V \to \mathbb{R} \cup \{\infty\}$ as

$$(4.3) \qquad \pi(\alpha z_t) := \sup\{f(x) : x \in P, g(x) + \alpha(0,\ldots,0,z_t,0,\ldots) \in Q\} .$$

Then, under conditions on the problem data of (RP) such that the original problem has an optimal solution, the perturbed problems $P[\alpha z_t]$ will have feasible solutions. We shall assume that we may find a curve $x(\alpha)$ such that $x(\alpha)$ is feasible for $P[\alpha z_t]$ and $\lim_{\alpha \downarrow 0} x(\alpha) = x_0 \in U$. Then, since the closed projection $(I - \Pi_t)$ defines a subspace of L_∞, the Lagrange multiplier $\rho_t' \in L_1^{n'}$ for the constraint (2.3) is an anihilator (supporting hyperplane) of this subspace. Under our assumptions (applying Theorem 4.3, Lempio and Maurer, 1980) we may thus conclude that we may choose

$$(4.4) \qquad \rho_t' = \nabla_t \pi(0) \quad ,$$

where $\nabla_t \pi$ denotes the Fréchet derivative of the perturbation function (4.3) of the abstract problem (P) evaluated at 0 under perturbations of the form (4.2) at time t. That is, the current state $\underset{\sim}{\rho}_t'$ of the nonanticipative constraint multiplier process $\underset{\sim}{\rho}_t'$ in $L_1^{n'}$ represents the *marginal expected value of perfect information (EVPI)* at time t with respect to future states of the observation process $\underset{\sim}{\xi}$.

We first establish that this marginal EVPI process $\underset{\sim}{\rho}'$ --like the optimal policy process $\underset{\sim}{x}_0$ itself--is adapted to the observation process $\underset{\sim}{\xi}$.

Lemma: 4.1. The process $\underset{\sim}{\rho}'$ in $L_1^{n'}$ is *nonanticipative*, i.e.

$$(4.5) \qquad \underset{\sim}{\rho}_t' = E\{\underset{\sim}{\rho}_t' | \Sigma_{t-1}\} \qquad \text{a.s.} \qquad t=1,2,\ldots \quad .$$

■ This fact follows from the observation that expression (4.4) for ρ_t' does not depend on any particular perturbation (4.2) representing some future knowledge of the observation process $\underset{\sim}{\xi}$. ■

Next we show that the process $\underset{\sim}{\rho}'$ has the *supermartingale property*. This reflects the fact that the earlier information on the future observation process $\underset{\sim}{\xi}$ is available, the more its marginal expected worth to optimal decision making.

Theorem: 4.2. The process $\underset{\sim}{\rho}'$ in L_n^1, is a *supermartingale*, i.e.

$$(4.6) \qquad \underset{\sim}{\rho}_t' \geqq E\{\rho_s' | \Sigma_t\} \qquad \text{a.s.} \qquad \text{for } (1 \leqq)\ t < s \ .$$

■ By virtue of (4.5) we must show for fixed t and $s > t$ that

$$E\{\underset{\sim}{\rho}_t' | \Sigma_{t-1}\} \geqq E\{\underset{\sim}{\rho}_s' | \Sigma_t\} \ \text{a.s.} \qquad .$$

But a further consequence of (4.5) is that for all $s \geq t$

$$E\{\underset{\sim}{\rho}_t' | \Sigma_s\} = E\{\underset{\sim}{\rho}_t' | \Sigma_{t-1}\} \ \text{a.s.} \qquad ,$$

and hence (4.6) is equivalent to showing that

$$E\{\underset{\sim}{\rho}_t' | \Sigma_t\} \geqq E\{\underset{\sim}{\rho}_s' | \Sigma_t\} \quad \text{a.s.} \qquad .$$

But information on the future of $\underset{\sim}{\xi}$ after time s-1, as represented by an n-vector valued perturbation function z measurable with respect to Σ_u for $u \geq s$, cannot be worth less in expectation the earlier it is known, i.e.

$$(4.7) \qquad \pi(\alpha z_t) \geq \pi(\alpha z_s) \quad ,$$

where $z_s := z_t := z$. Indeed, an optimal policy for the problem perturbed at time t can take this information into account earlier than a corresponding policy for the problem perturbed at s . Hence, subtracting $\pi(0)$ from each side of (4.7), dividing by $\alpha > 0$ and passing to the limit as $\alpha \rightarrow 0$, yields

$$\rho_t' = \nabla_t \pi(0) \geq \nabla_s \pi(0) = \rho_s'$$

Since integration is nonnegativity preserving

$$E\{\underset{\sim}{\rho}_t | \Sigma_t\} \geq E\{\underset{\sim}{\rho}_s | \Sigma_t\} \qquad \text{a.s.} \quad . \qquad \blacksquare$$

5. POSSIBLE EXTENSIONS

As noted in the introduction, the marginal expected value of perfect information (EVPI) process $\underset{\sim}{\rho}'$ is of considerable *potential* importance for stochastic systems of the dynamic recourse type arising in practice (see §1). If this nonanticipative supermartingale process in $L_1^{n'} := L_1[(\Xi,\Sigma,\mu); \mathbb{R}^{n'}]$ remains in a ball of (problem dependent) radius $\varepsilon > 0$ for all t after some time $s \geq 1$, then the stochastic elements of the problem are practically inessential from time s onward and a *deterministic* model--and *simpler* computational procedure--should suffice. Of course, this statement raises the knotty problems of prior numerical computation of the marginal EVPI process, or--more realistically--of bounds on this process, etc. (in this context, see Birge,1980).

Nevertheless, it would be interesting to have theoretical results similar to those derived in §4 for familiar optimization of stochastic system problems in *continuous time* involving dynamics driven by *semimartingales* (see e.g. Shiryaev,1980). The difficulty in attempting an analogue of the analysis presented in this paper for such systems is that the corresponding perturbed abstract problem (as utilized in §4) must make sense. Put differently, the original stochastic optimization problem must remain well defined when nonanticipativity is relaxed. Using the Ito calculus approach (and its recent extensions to semimartingales generating mixed diffusion and jump dynamics) this is not possible, since the rigorous analytic *integral* form of the dynamics requires nonanticipativity of the integrand in the stochastic integrals involved. This *technical* requirement of the stochastic integration theories utilized has been relaxed for integration of *Gaussian* processes with respect to similar processes by Enchev and Stoyanov (1980), but this setting is of insufficient generality for many systems of interest. More promising is the application to the problem at hand of the recent *pathwise* theory of stochastic integration introduced for the study of stochastic differential equations whose integrals are driven by processes with continuous sample paths by Sussman (1978) and developed for semimartingales with jumps, for example, by Marcus (1981).

In the case of successful application of the approach of this paper to optimization of stochastic systems in continuous time, with differential dynamics in $\mathbb{R}_n$ of the form $\dot{\underset{\sim}{x}} = \underset{\sim}{f}(\underset{\sim}{x})$, it may be conjectured that the *full* expected value of perfect information process $\underset{\sim}{\sigma}$ in L_1 may be recovered from the *marginal* EVPI process $\underset{\sim}{\rho}'$ in $L_1^{n'}$ by (Lebesque) integration as

(5.1) $$\underset{\sim}{\sigma}_t = \int_t^\tau \underset{\sim}{\rho}'_s \, \dot{\underset{\sim}{x}}_s \, ds$$

for an appropriate definition of $\dot{\underset{\sim}{x}}_s$. This is again a statement of some potential practical importance for stochastic system modelling.

6. ACKNOWLEDGEMENTS

I would like to express my gratitude to J-M. Bismut, who first pointed out to me the technical difficulties discussed in §4 regarding extension of the present analysis to stochastic systems in continuous time, and to M.H.A. Davis, whose conversation lead to the suggestions made therein for surmounting them.

REFERENCES

1. J. Birge (1981). Stochastic dynamic linear programs. Ph.D. Thesis Operations Research Department, Stanford University.

2. M. Brokate (1980). A regularity condition for optimization in Banach spaces: Counter examples. *Appl.Math.Optim. 6*, 189-192.

3. M.A.H. Dempster (1976). Lectures on Abstract Optimization and Its Applications. Department of Mathematics, Melbourne University.

4. M.A.H. Dempster (1980). Introduction to stochastic programming. In: *Stochastic Programming*. M.A.H. Dempster, ed. Academic, London. 3-59.

5. A.Y. Dubovitskii & A.A. Milyutin (1965). Extremum problems in the presence of restrictions. *USSR Comput.Math. & Math.Phys. 5.3*, 1-80.

6. N. Dunford & J.J. Schwartz (1956). *Linear Operators*. Vol.I. Interscience, New York.

7. M.J. Eisner & P. Olsen (1975). Duality for stochastic programming interpreted as L.P. in L_p-space. *SIAM J.Appl.Math. 28*, 779-792.

8. M.J. Eisner & P. Olsen (1980). Duality in probabilistic programming. In: *Stochastic Programming, op.cit.*, 147-158.

9. O.B. Enchev & J.M. Stoyanov (1980). Stochastic integrals for Gaussian random functions. *Stochastics 3*, 277-289.

10. R.C. Grinold (1976). Manpower planning with uncertain demands. *Operations Res. 24*, 387-399.

11. R.C. Grinold (1980). A class of constrained linear control problems with stochastic coefficients. In: *Stochastic Programming, op.cit.*, 97-108.

12. J-B. Hiriart-Urruty (1978). Conditions necessaires d'optimalité pour un programme stochastique avec recours. *SIAM J. Control Optim. 16.2*, 317-329.

13. J-B. Hiriart-Urruty (1981). Extension of Lipschitz integrands and minimization of nonconvex integral functionals: Applications to the optimal recourse problem in discrete time. *Probability and Statistics 1*, to appear.

14. F. Lempio & H. Maurer (1980). Differential stability in infinite-dimensional nonlinear programming. *Appl.Math.Optim. 6*, 139-152.

15. F. Louveaux & Y. Smeers (1981a). A stochastic model for electricity generation. In: *Modelling of Large-Scale Energy Systems*. W. Haefele & L.K. Kirchmayer, eds. IIASA Proc.Series No.12. Pergamon, Oxford, 313-320.

16. F. Louveaux & Y. Smeers (1981b). Optimization of a stochastic model for electricity generation. *Stochastics*. Special issue containing the Proceedings of the IIASA Task Force Meeting on Stochastic Optimization, 9-13 December, 1980. To appear.

17. S.I. Marcus (1981). Modelling and approximation of stochastic differential equations driven by semimartingales. *Stochastics 4.3*, 223-246.

18. E.C. Prescott & R.E. Lucas, Jr. (1972). A note on price systems in infinite dimensional space. *Intl.Econ.Rev. 13*, 416-422.

19. R.T. Rockafellar & R.J-B. Wets (1976a). Stochastic convex programming: Relatively complete recourse and induced feasibility. *SIAM J.Control Optim. 14*, 574-589.

20. R.T. Rockafellar & R.J-B. Wets (1976b). Nonanticipativity and L^1-martingales in stochastic optimization problems. *Math.Programming Studies 6*, 170-187.

21. R.T. Rockafellar & R.J-B. Wets (1978). The optimal recourse problem in discrete time: L^1-multipliers for inequality constraints. *SIAM J.Control Optim. 16.1*, 16-36.

22. R.T. Rockafellar & R.J-B. Wets (1981). Stochastic optimization problems of Bolza type in discrete time. *Stochastics*. IIASA Special Issue, *op.cit.* To appear.

23. A.N. Shiryayev (1980). Martingales: Recent Developments, Results, Applications. Lecture notes distributed at the 1980 European Meeting of Statisticians, Brighton, England.

24. H.J. Sussmann (1978). On the gap between deterministic and stochastic ordinary differential equations. *Ann.Prob. 6.1*, 19-41.

25. M. Valadier (1974). A natural supplement of L_1 in the dual of L_∞. Seminaire d'Analyse Convexe, Montpelier.

26. A.F. Veinott (1966). The status of mathematical inventory theory. *Management Sci. 12*, 745-777.

27. K. Yosida & E. Hewitt (1952). Finitely additive measures. *Trans. A.M.S. 72*, 46-66.

28. J. Zowe & S. Kurcyusz (1979). Regularity and stability for the mathematical programming problem in Banach spaces. *Appl.Math. Optim. 5*, 49-62.

SOME PROBLEMS OF LARGE DEVIATIONS

M.D. Donsker and S.R.S. Varadhan
Courant Institute of Mathematical Sciences
New York University
New York, NY 10012/USA

Let E_t refer to the expectation with respect to a three dimensional Brownian path $\beta(\cdot)$ tied down at both ends with $\beta(0) = \beta(t) = 0$. Let

$$G(\alpha,t) = E_t\left\{\exp\left[\alpha \int_0^t\int_0^t \frac{e^{-|\sigma-s|}}{|\beta(\sigma)-\beta(s)|}\, d\sigma\, ds\right]\right\}$$

show that

$$\lim_{t\to\infty} \frac{1}{t} \log G(\alpha,t) = g(\alpha) \quad \text{exists}$$

and

$$\lim_{\alpha\to\infty} \frac{g(\alpha)}{\alpha^2} = g_0 \quad \text{exists}$$

with

$$g_0 = \sup_{\substack{\phi\in L_2(R^3)\\ \|\phi\|_2=1}} \left[2\int\int \frac{\phi^2(x)\phi^2(y)}{|x-y|}\, dx\, dy - \frac{1}{2}\int |\nabla\phi|^2\, dx\right]$$

The problem comes up in stastical mechanics. See for instance the book by Feynman [3]. The formula for g_0 has been conjectured by Pekar [4]. We shall outline a theory that allows us to prove these formulae.

REDUCTION 1

One can replace the tied down Brownian by the free Brownian motion that starts at time 0 from 0. This is easily justified because for large t the Brownian motion for the most part does not feel the condition at the terminal time t.

REDUCTION 2

$$\int_0^t\int_0^t \frac{e^{-(\sigma-s)}}{|\beta(\sigma)-\beta(s)|}\, d\sigma\, ds = 2\int_0^t\int_s^t \frac{e^{-(\sigma-s)}}{|\beta(\sigma)-\beta(s)|}\, d\sigma\, ds$$

$$\simeq 2 \int_0^t ds \int_s^\infty \frac{e^{-(\sigma-s)}}{|\beta(\sigma)-\beta(s)|} d\sigma$$

$$= 2 \int_0^t ds \int_s^\infty \frac{e^{-\tau} \, d\tau}{|\beta(s+\tau)-\beta(s)|}$$

$$= 2 \int_0^t F(\omega_s) \, ds$$

where

$$F(\omega) = 2\alpha \int_0^\infty \frac{e^{-\tau} \, d\tau}{|\beta(\tau)-\beta(0)|}, \quad \omega = \beta(\cdot), \ \omega_s = \beta(\cdot+s)$$

The first question then becomes the evaluation of

$$\lim_{t\to\infty} \frac{1}{t} \log E\left[\int_0^t F(\omega_s) \, ds \right]$$

where E refers to the expectation with respect to the three dimensional Brownian motion starting at the origin or more generally any homogeneous Markov process.

Let Q be any stationary stochastic process with values in R^3 and let F_- be the σ-field generated by the process $x(t)$ for $t \leq 0$. We denote by F_T^0 the σ-field generated by the process $x(t)$ for $0 \leq t \leq T$. It is convenient to take the space Ω of continuous trajectories as the basic space and view F_- and F_T^0 as σ-fields of Ω, in fact sub-σ-fields of F the entire σ-field on which Q is a translation invariant measure.

Let Q_ω^T be the regular conditional probability of Q on F_T^0 given F_-. We denote by $P_{\omega(0)}^T$ the measure on F_T^0 corresponding to the Brownian motion starting at time 0 from $\omega(0)$. Let $h_T(\omega)$ be the entropy

$$h_T(\omega) = \int \log R_\omega^T(\omega') \, Q_\omega^T(d\omega')$$

where $R_\omega^T(\cdot)$ is the Radon-Nikodym derivative of Q_ω^T with respect to $P_{\omega(0)}^T$ on F_T^0. The function $h_T(\omega)$ is always well defined and $0 \leq h_T(\omega) \leq \infty$. When it is not naturally defined it can be taken to be $+\infty$. We then define

$$h(T) = \int h_T(\omega) \, dQ$$

It then turns out that either $h(T) = \infty$ for all $T > 0$ or $h(T) = h\cdot_{\perp}$ where for some constant h which depends on Q; we can therefore write

$$h(T) = T\,H(Q) ,$$

for some $0 \leq H(Q) \leq \infty$. We then have the following theorem which is valid under suitable assumptions.

Theorem 1

$$\lim_{t\to\infty} \frac{1}{t} \log E\left\{\exp\left[\int_0^t F(\omega_s)\, ds\right]\right\} = \sup_Q \left\{E^Q[F(\omega)] - H(Q)\right\}$$

where the supremum is taken over all stationary processes.

One can verify that the theorem applies to our example so that $g(\alpha)$ exists and is given by

$$g(\alpha) = \sup_Q \left[E^Q\left\{2\alpha \int_0^\infty \frac{e^{-t}\, dt}{|x(t) - x(0)|}\right\} - H(Q)\right]$$

Using Brownian scaling one can write

$$g(\alpha) = \sup_Q \left[E^Q\left\{2\alpha \int_0^\infty \frac{e^{-t}\, dt}{|x(t) - x(0)|} - H(Q)\right]\right.$$

$$= \alpha^2 \sup_Q \left[E^Q\left\{2 \int_0^\infty \frac{e^{-t}\, dt}{|x(\frac{t}{\alpha^2}) - x(0)|}\right\} - H(Q)\right]$$

The next problem then is to evaluate

$$g_0 = \lim_{\alpha\to\infty} \frac{g(\alpha)}{\alpha^2} = \lim_{\alpha\to\infty} \sup_Q \left[E^Q\left\{2 \int_0^\infty \frac{e^{-t}\, dt}{|x(\frac{t}{\alpha^2}) - x(0)|}\right\} - H(Q)\right]$$

One notes that $H(Q)$ is also linear in Q and the supremum is therefore attained at an extremal which is an ergodic process. Interchanging the supremum and the limit which can be justified we find that by the ergodic theorem

$$\lim_{\alpha\to\infty} E^Q\left\{2 \int_0^\infty \frac{e^{-t}\, dt}{|x(\frac{t}{\alpha^2}) - x(0)|}\right\} = \iint \frac{2}{|x(0) - y(0)|}\, Q(dx)\, Q(dy)$$

where x,y are two independent versions. Denoting by λ_Q the marginal distribution of Q at any time we can see that

$$\lim_{\alpha\to\infty} \frac{g(\alpha)}{\alpha^2} = g_0 = \sup_Q \left[2 \iint \frac{\lambda_Q(dx)\ \lambda_Q(dy)}{|x-y|} - H(Q) \right]$$

$$= \sup_\lambda \left[2 \iint \frac{\lambda(dx)\ \lambda(dy)}{|x-y|} - I(\lambda) \right]$$

where

$$I(\lambda) = \inf_{Q:\lambda_Q=\lambda} H(Q)$$

It can be shown that $I(\lambda) < \infty$ if and only if $\lambda(dx) = \phi^2(x)\,dx$ for some $\phi \in L^2(R^3)$ with $\|\phi\|_2 = 1$ and $\int |\nabla\phi|^2\,dx < \infty$. In such a case

(*) $$I(\lambda) = \frac{1}{2} \int |\nabla\phi|^2\,dx\ .$$

We have now established Pekar's formula

$$g_0 = \sup_{\substack{\phi\in L_2(R^3) \\ \|\phi\|_2 = 1}} \left[2 \iint \frac{\phi^2(x)\ \phi^2(y)}{|x-y|}\,dx\,dy - \frac{1}{2} \int |\nabla\phi|^2\,dx \right]$$

We shall explain in the next few pages the rationale behind Theorem 1, and formula (*).

Suppose $x_1, x_2, \ldots, x_n, \ldots$ are independent identically distributed random variables with a moment generating function

$$M(\theta) = \int e^{\theta x}\,dF(x)$$

where F is the common distribution of x_i. Assume that $Ex_i = 0$. Then Cramer [1] showed that for any $a > 0$,

$$P\left\{\frac{x_1 + \cdots + x_n}{n} \geq a\right\} = \exp\left\{- n\,h(a) + o(n)\right\}$$

as $n \to \infty$ where

$$h(a) = \sup_\theta\,[\theta a - \log M(\theta)]$$

We can look at the problem at a higher level if we consider the sample distribution function $(\delta x_1+\ldots+\delta x_n)/n$ as a random measure and ask about $P[(\delta x_1+\ldots+\delta x_n)/n \in N_\mu]$ where N_μ is some "tiny" neighborhood of a measure μ which is different from μ. One can show that

$$P\left[\frac{\delta x_1 + \ldots + \delta x_n}{n} \in N_\mu\right] = \exp\left\{- nI_F(\mu) + o(n)\right\}$$

where

$$I_F(\mu) = \int \log\left(\frac{d\mu}{dF}\right)(x)\; d\mu(x)$$

is the entropy. Since the mean of the sample distribution is the sample mean compatibility with Cramer's result implies that

$$h(a) = \inf_{\mu:\int x\, d\mu = a} I_F(\mu)$$

We can even go one step higher by starting from $x_1,\ldots,x_n$ and creating a doubly infinite periodic sequence

$$\xi = (\;\ldots\; x_1 \;\ldots\; x_n \; x_1 \;\ldots x_n \;\ldots\;)$$

We view $(\delta_\xi+\delta_{T\xi}+ \ldots + \delta_{T^{n-1}\xi})/n$ as a random stationary stochastic process and call it $R_{n,\omega}$. If we denote by P the product measure based on F we can ask for the behavior of

$$P\left\{R_{n,\omega} \in N_Q\right\}$$

where N_Q is a "tiny" neighborhood of a stationary process Q. We can show

$$P\left\{R_{n,\omega} \in N_Q\right\} \simeq \exp\left[-n\, I_P(Q) + o(n)\right]$$

where $I_P(Q)$ is analogous to Shannon entropy. If Q_ω is the regular conditional probability X_0 given $x_{-1},\ldots,x_{-n},\ldots$ under Q then

$$I_P(Q) = \int I_F(Q_\omega)\; dQ$$

Again since the sample distribution function of $x_1,\ldots,x_n$ is the marginal of $R_{n,\omega}$ if we denote by λ_Q the one dimensional marginal of a stationary process Q we must have

$$\inf_{Q:\lambda_Q=\lambda} I_P(Q) = I_F(\lambda)\;.$$

For our application we must seek natural generalizations to the case where $x_1, \dots, x_n, \dots$ form a Markov process instead of being independent and replace discrete time by continuous time.

Finally, once we have large deviation probabilities we can evaluate the integrals asymptotically by Laplace's method. See in this connection [1].

Acknowledgement

This work was supported in part by the National Science Foundation, under Grant No. MCS 80 02568.

REFERENCES

1. Cramér, H. On a new limit theorem in the theory of probability, Colloquium on the Theory of Probability, Hermann Paris, 1937.
2. Donsker, M. D. and Varadhan, S. R. S. Asymptotic evaluation of certain Markov process expectations for large time, I, II, III, Comm. Pure Appl. Math. 28 (1975) 1-47; 28 (1975) 279-301; 30 (1976), 389-461.
3. Feynman, R. P. Statistical Mechanics, W. A. Benjamin, 1972.
4. Pekar, S. I., Theory of Polarons, Zh. Eksperim. i Teor. Fiz., v. 19, 1949.

ON THE BEHAVIOUR OF CERTAIN FUNCTIONALS OF THE WIENER PROCESS AND APPLICATIONS TO STOCHASTIC DIFFERENTIAL EQUATIONS

H.J. Engelbert and W. Schmidt
University of Jena
GDR

The purpose of this paper is to investigate the convergence of certain functionals of the one-dimensional Wiener process. In section I (Theorem 1) we prove a 0-1 law and, furthermore, we give necessary and sufficient conditions for the convergence of these functionals.
In a second section the results are applied to the study of some functionals of the Wiener process associated with stochastic differential equations. Finally, we formulate an analogous result for process that are solutions of stochastic differential equations up to the first exit time of an interval.
The main results are Theorem 1 and Theorem 4. They are presented with detailed proofs whereas the proofs of some other statements will be published elsewhere.
In the following by (X,F) we denote a real valued stochastic process $(X_t)_{t\geq 0}$ on a probability space $(\Omega,\underline{\underline{F}},P)$ where $F=(\underline{\underline{F}}_t)_{t\geq 0}$ is an increasing family of sub-σ-algebras of $\underline{\underline{F}}$ and X_t is $\underline{\underline{F}}_t$-measurable for all $t\geq 0$.

I. A 0-1 LAW

Let (W,F) be a Wiener process on a probability space $(\Omega,\underline{\underline{F}},P)$.

THEOREM 1. Suppose that f is a Borel measurable function of the real line into $[0,\infty]$. Then the following assertions are equivalent.

(i) $P(\{\int_0^t f(W_s)ds < \infty$, for every $t\geq 0\}) > 0$

(ii) $P(\{\int_0^t f(W_s)ds < \infty$, for every $t\geq 0\}) = 1$

(iii) $\int_K f(y)dy < \infty$ for all compact subsets K of the real line.

REMARK 1. Each of the following three conditions is also equivalent to the assertions in Theorem 1.

(iv) There exists a $t_o > 0$ such that

$$P(\{ \int_0^{t_o} f(W_s)ds < \infty \}) = 1.$$

(v) For every $x_o \in (-\infty, \infty)$

$$P(\{ \int_0^t f(x_o+W_s)ds < \infty \text{ , for every } t \geq 0 \}) = 1.$$

(vi) For every $x_o \in (-\infty, \infty)$ there exists a Wiener process (W^{x_o}, F^{x_o}) on a probability space $(\Omega^{x_o}, \underline{\underline{F}}^{x_o}, P^{x_o})$ and a random time τ, strictly positive and measurable, such that

$$P(\{ \int_0^{\tau} f(x_o+W_s^{x_o})ds < \infty \}) > 0.$$

It turns out that the condition $P(\{ \int_0^{t_o} f(W_s)ds < \infty \}) > 0$ for some $t_o > 0$ is not sufficient for (iv) in contrast to the sufficiency of (i) for (ii) in Theorem 1.

A 0-1 law of the form $P(\{ \int_0^{t_o} f(W_s)ds < \infty \}) \in \{0,1\}$ where the nonnegative function f is bounded on $\{x: |x| > \varepsilon\}$ for every $\varepsilon > 0$ was proved by L.A. Shepp, J.R. Klauder and H. Ezawa /7/ in 1974.

However, as above noted, a 0-1 law of that type does not hold for general functions f.

The basic idea of the proof of Theorem 1 consists in the use of the local time of a Wiener process (cf. /7/).

First we prove the following lemma.

LEMMA. If for a point $x_o \in (-\infty, \infty)$ there exists a random time τ, strictly positive and measurable, such that $P(\{ \int_0^{\tau} f(x_o+W_s)ds < \infty \}) > 0$ then there is a neighbourhood U of x_o with $\int_U f(y)dy < \infty$.

PROOF. Set $A = \{ \int_0^{\tau} f(x_o+W_s)ds < \infty \}$. We denote by $\mathcal{A}$ the local time of the Wiener process (W,F) (cf. K. Ito, H.P. McKean Jr. /5/) wich is defined as a real function on $(-\infty, \infty) \times [0, \infty) \times \Omega$ with

$$\text{Lebesgue meas.}(\{ s \leq t: W_s(\omega) \in B \}) = \int_B \mathcal{A}(y,t,\omega)dy, (t,\omega) \in [0,\infty) \times \Omega ,$$

for each Borel subset B of $(-\infty, \infty)$ and such that it is continuous in (y,t) P-a.s. .

Then we have

$$\int_0^{\tau(\omega)} f(x_o+W_s(\omega))ds = \int_{-\infty}^{\infty} f(x_o+y) \mathcal{A}(y,\tau(\omega),\omega)dy \qquad (*).$$

We recall some properties of the local time $\mathcal{A}$ we want to use.

First it is clear from the definition that for P-allmost all $\omega \in \Omega$

the local time $\mathcal{A}(y,t,\omega)$ is increasing in t for all $y \in (-\infty,\infty)$.
It is also well-known that

$$P(\{ \mathcal{A}(0,s,.) > 0 \}) = 1 \qquad \text{for all } s > 0$$

(cf. /5/, p. 71) and therefore, because of the above mentioned monotony of $\mathcal{A}$ in t,

$$P(\{ \mathcal{A}(0,s,.) > 0, \text{ for all } s \geq \tfrac{1}{n} \}) = 1 \qquad \text{for all } n=1,2,3,\dots .$$

From this it follows that the set

$$\{\mathcal{A}(0,s,.) > 0, \text{ for all } s > 0\} = \bigcap_n \{\mathcal{A}(0,s,.) > 0, \text{ for all } s \geq \tfrac{1}{n}\}$$

has probability one and thus

$$P(\{ \mathcal{A}(0, \tau(.),.) > 0 \}) = 1.$$

Now the continuity of $\mathcal{A}(y, \tau(.),.)$ in y P-a.s. implies that for P-allmost all $\omega \in \Omega$ there exists a neighbourhood $U(\omega)$ of the point 0 and a strictly positive constant $c(\omega)$ such that

$$\mathcal{A}(y,\tau(\omega),\omega) \geq c(\omega) > 0 \quad \text{for all } y \in U(\omega).$$

Using this we obtain the inequality

$$\int_{-\infty}^{\infty} f(x_o+y)\, \mathcal{A}(y,\tau(\omega),\omega)dy \geq c(\omega) \int_{U(\omega)} f(x_o+y)dy \qquad \text{P-a.s. .}$$

In view of (*) the left-hand side is finite P-a.s. on A. By assumption, we have $P(A) > 0$ and the assertion follows.┘

Now we come to the proof of Theorem 1.

PROOF OF THEOREM 1. Let (i) be fulfiled. Our aim is to show that for each $x_o \in (-\infty,\infty)$ the assumption of the Lemma is valid.
Let $x_o \in (-\infty,\infty)$ and $t > 0$ be arbitrary but fixed and define
$\tau_{x_o} = \inf\{s \geq 0 : W_s = x_o\}$.
Then it follows from (i) that $P(\{ \int_0^{\tau_{x_o}+t} f(W_s)ds < \infty \}) > 0$.
Because $\tilde{W}_s = W_{\tau_{x_o}+s} - x_o$ for $s \geq 0$ is again a Wiener process and
$P(\{ \int_0^t f(x_o+\tilde{W}_s)ds < \infty \}) > 0$ in view of $\int_0^t f(x_o+\tilde{W}_s)ds \leq \int_0^{\tau_{x_o}+t} f(W_s)ds$ the assumption of the Lemma is fulfilled for $\tilde{W}$.
Thus for each $x_o \in (-\infty,\infty)$ there exists an open neighbourhood U_{x_o} of the point x_o with $\int_{U_{x_o}} f(y)dy < \infty$.

Now let K be a compact subset of $(-\infty,\infty)$. The system $(U_{x_o})_{x_o \in K}$ of open sets covers K. Hence this system contains a finite subsystem which also covers K and therefore

$$\int_K f(y)dy < \infty \quad .$$

Because K is arbitrary we obtain (iii).
Next suppose that (iii) is satisfied.
As in the proof of the Lemma we can write

$$\int_0^t f(W_s(\omega))ds = \int_{-\infty}^{\infty} f(y) A(y,t,\omega)dy \qquad P\text{-a.s.}\ .$$

By M_t and m_t denote the maximum and minimum of the sample path of our given Wiener process W up to time t. Since $A(y,t,\omega)$ vanishes for $y \notin [m_t(\omega), M_t(\omega)]$ we have

$$\int_0^t f(W_s(\omega))ds = \int_{m_t(\omega)}^{M_t(\omega)} f(y) A(y,t,\omega)dy$$

$$\leq \max_{y\in[m_t(\omega),M_t(\omega)]} A(y,t,\omega) \int_{m_t(\omega)}^{M_t(\omega)} f(y)dy \qquad P\text{-a.s.}\ .$$

But this is less than infinity P-a.s. because of (iii) and the continuity of the local time A in y.┘

The conditions in Remark 1 are consequences of the proof of Theorem 1. Indeed, if (vi) is fulfilled we can apply the Lemma for each $x_o \in (-\infty,\infty)$ and obtain (iii) as in the first part of the proof of Theorem 1. If (iii) is satisfied for f then (iii) is also satisfied for the function f_{x_o} with $f_{x_o}(y) = f(x_o+y), y \in (-\infty,+\infty)$. By Theorem 1, the condition (ii) for f_{x_o} holds and, consequently, (v) is satisfied. It is obvious that (v) implies (vi) and that (iv) follows from (ii). Finally, let (iv) be true. Then for $\tau_{x_o} = \inf\{s\geq 0 : W_s = x_o\}$ and $0 < \tau < t_o$ we have $P(\{\tau_{x_o} + \tau \leq t_o\}) > 0$ for every $x_o \in (-\infty,\infty)$. From (iv) it follows for P-allmost all $\omega \in \{\tau_{x_o} + \tau \leq t_o\}$ that

$$\int_0^{\tau} f(x_o+W_s^{x_o})ds \leq \int_0^{\tau_{x_o}+\tau} f(W_s)ds \leq \int_0^{t_o} f(W_s)ds$$

where $W_s^{x_o} = W_{\tau_{x_o}+s} - x_o$ is again a Wiener process.

But this means that (vi) is fulfiled.

Theorem 1 can be generalized in the following manner. Let (u,v) be an interval of the real line and by $\tau^W_{(u,v)}$ denote the first exit time of (W,F) from (u,v):

$$\tau^W_{(u,v)} = \inf\{s\geq 0 : W_s \notin (u,v)\}\ .$$

THEOREM 2. Let f be as in Theorem 1. The following conditions are equivalent.

(i) $P(\{\int_0^t f(W_s)ds < \infty \ ,\text{ for every } t < \tau^W_{(u,v)}\}) > 0$

(ii) $P(\{\int_0^t f(W_s)ds < \infty \ ,\text{ for every } t < \tau^W_{(u,v)}\}) = 1$

(iii) $\int_K f(y)dy < \infty$ for all compact subsets K of the interval (u,v).

Obviously, Theorem 1 is obtained by setting $u=-\infty$, $v=+\infty$.

Clearly $P(\{\int_0^\infty f(W_s)ds < \infty\}) = 1$ need not hold if (i) in Theorem 1 is true. For example, let f be the function that is identically equal to 1. The following theorem even shows that if $P(\{\int_0^\infty f(W_s)ds < \infty\}) > 0$ then the function f can only be different from zero on a set of Lebesgue measure zero.

THEOREM 3. If the nonnegative Borel measurable real function f satisfies

Lebesgue meas. $(\{y : f(y) > 0\}) > 0$

then $P(\{\int_0^\infty f(W_s)ds = \infty\}) = 1$.

II. APPLICATIONS TO STOCHASTIC DIFFERENTIAL EQUATIONS

We consider the one-dimensional stochastic differential equation

$$dX_t = a(X_t)dt + b(X_t)dW_t^* \ , \ t \geq 0 \qquad (1)$$

where W^* is a Wiener process and a, b are real Borel measurable functions.

A stochastic process (X,F) defined on a certain probability space $(\Omega,\underline{\underline{F}},P)$ is called a <u>weak solution of (1)</u>, with the initial distribution P_o, if there exists a Wiener process (W^*,F) on the same probability space such that (1) and $P(\{X_o \in B\}) = P_o(B)$ for all Borel subsets B of the real line hold.

It turns out that the solution of (1) is reducible to the solution of

$$dX_t = b(X_t)dW_t^* \ , \ t \geq 0 \ . \qquad (2)$$

Under certain conditions the general equation can then be solved by the well-known theorem of Girsanov.

In Theorem 4 we formulate a necessary and sufficient condition for the existence of a weak solution of equation (2). For simplicity we restrict ourselves to deterministic initial conditions $P(\{X_o = x_o\}) = 1$.

A process (X,F) is said to be <u>trivial</u> if

$P(\{X_t = X_o, \text{ for all } t \geq 0\}) = 1$.

THEOREM 4. The following assertions are equivalent.

(i) For every $x_o \in (-\infty,\infty)$ there exists a nontrivial weak solution (X,F) of equation (2) with $X_o \equiv x_o$

(ii) $\int_K b^{-2}(y)dy < \infty$ for all compact subsets K of the real line (with the convention $b^{-2}(y) = \infty$ if $b(y) = 0$).

PROOF. We use the method of random time change to construct a solution of (2) starting from a Wiener process (W,F^W) on a certain probability

space $(\Omega,\underline{\underline{F}},P)$.
This method was already used by I.I. Gichman and A.V. Skorochod /4/, M.P. Yershov /9/ and S. Watanabe /8/ to find sufficient conditions for the existence of a weak solution of (2). But they assumed either the continuity of b, or its boundedness, or the boundedness of b^2 away from zero.
Let (ii) be fulfiled. We suppose that $\underline{\underline{F}}$ is complete with respect to P and that each $\underline{\underline{F}}_t^W$ of the right continuous family F^W contains all $\underline{\underline{F}}$-sets of probability zero. Define the continuous, strictly increasing F^W-adapted process T by

$$T_t = \int_0^t b^{-2}(x_o+W_s)ds, \quad t\geq 0.$$

From Remark 1 (v) it follows that $T_t<\infty$ for all $t\geq 0$ P-a.s. .
An application of Theorem 3 shows that $\lim_{t\to\infty} T_t = T_\infty = \infty$ P-a.s. .
Now the right inverse of (T,F^W)

$$A_t = \inf\{s\geq 0 : T_s > t\},\ t\geq 0$$

is a P-a.s. strictly increasing continuous family $(A_t)_{t\geq 0}$ of finite F^W-stopping times. Hence the process $X_t=x_o+W_{A_t}$, $t\geq 0$ is a continuous local martingale with respect to $F=(\underline{\underline{F}}_t)_{t\geq 0}$ where $\underline{\underline{F}}_t = \underline{\underline{F}}_{A_t}^W$ for $t\geq 0$ and the associated increasing previsible process from the Doob-Meyer decomposition of (X^2,F) is

$$\langle X\rangle_t = A_t\ ,\quad t\geq 0$$

(cf. N. Kazamaki /6/).
Using the properties of T and that the sample functions $A.(\omega)$ are P-a.s. strictly increasing it follows

$$\langle X\rangle_t = \int_0^{A_t} b^2(x_o+W_s)dT_s = \int_0^{T_\infty\wedge t} b^2(x_o+W_{A_s})ds = \int_0^t b^2(X_s)ds \quad ,\ t\geq 0$$

(for the transformation of the integral see C. Dellacherie /1/,T IV.43)
Because of a well-known theorem of Doob there exists a Wiener process (W^*,F) such that

$$X_t = X_o + \int_0^t b(X_s)dW_s^* \quad ,\ t\geq 0\ ,$$

i.e. (X,F) is a weak solution of (2) with the initial condition $X_o=x_o$.
Now let (i) be fulfilled. For an arbitrary but fixed $x_o\in(-\infty,\infty)$, by (X,F) denote a nontrivial solution defined on a complete probability space $(\Omega,\underline{\underline{F}},P)$ with $X_o=x_o$. We suppose that F is right continuous and

$\underline{F}_o$ contains all events from $\underline{F}$ having probability zero.
It is well-known that (X,F) is a continuous local martingale satisfying $\langle X\rangle_t = \int_0^t b^2(X_s)ds$. Consequently,

$$W_t = X_{T_t} - x_o , \quad t\geq 0 ,$$

where $T_t = \inf\{s\geq 0 : \langle X\rangle_s > t\}$, is a Wiener process stopped at $\langle X\rangle_\infty = \lim_{t\to\infty} \langle X\rangle_t$ and adapted to $F^W = (\underline{F}^W_t)_{t\geq 0}$ defined by $\underline{F}^W_t = \underline{F}_{T_t}$ for $t\geq 0$ (cf. H.J. Engelbert, J.Hess /2/).
Next we use that for $t\geq 0$

$$t \geq \int_0^t b^{-2}(X_s)\, b^2(X_s) I_{\{b^2(X_s)>0\}}(s)\, ds$$

$$= \int_0^t b^{-2}(X_s)\, I_{\{b^2(X_s)>0\}}(s)\, d\langle X\rangle_s \qquad \text{P-a.s.}$$

$$= \int_0^t b^{-2}(X_s) d\langle X\rangle_s \qquad \text{P-a.s.}$$

because of $d\langle X\rangle_s = b^2(X_s)ds$ and $\int_0^t I_{\{b^2(X_s)=0\}}(s)\, d\langle X\rangle_s = 0$.

In particular,

$$T_t \geq \int_0^{T_t} b^{-2}(X_s) d\langle X\rangle_s \qquad \text{P-a.s.}$$

and thus

$$T_t \geq \int_0^{\langle X\rangle_\infty \wedge t} b^{-2}(X_{T_s})ds = \int_0^{\langle X\rangle_\infty \wedge t} b^{-2}(x_o+W_s)ds, \qquad \text{P-a.s.} . \quad (3)$$

Now there exists a strictly positive random time τ such that $\tau(\omega) < \langle X\rangle_\infty(\omega)$ on the set $\{\langle X\rangle_\infty > 0\}$ (cf. C. Dellacherie /1/, T I.37) and consequently $T_{\tau(\omega)} < \infty$ on this set.
Because of the nontriviality of the solution we have $P(\{\langle X\rangle_\infty > 0\}) > 0$ and, therefore, it follows from (3)

$$P(\{\int_0^\tau b^{-2}(x_o+W_s)ds < \infty\}) > 0.$$

We can carry out the considerations above for every $x_o \in (-\infty,\infty)$ and thus condition (vi) of Remark 1 is satisfied. ⌟

Theorem 2 and the proof of Theorem 4 suggest a natural generalization of Theorem 4.
For this purpose we define the notion of a weak solution of the equation (1) on an interval.
Let (u,v) be an interval of the real line.

A stochastic process (X,F) defined on a certain probability space $(\Omega,\underline{F},P)$ is called a <u>weak solution of (1) on the interval (u,v)</u> if

there exists a Wiener process (W^*,F) on the same probability space such that (1) holds up to the first exit time $\tau^X_{(u,v)} = \inf\{s \geq 0 : X_s \notin (u,v)\}$ of the interval (u,v).
The following theorem generalizes Theorem 4.

THEOREM 5. The following conditions are equivalent.

(i) For every $x_0 \in (u,v)$ there exists a nontrivial weak solution (X,F) of the equation (2) on the interval (u,v) with $X_0 \equiv x_0$.

(ii) $\int_K b^{-2}(y)dy < \infty$ for all compact subsets K of the interval (u,v) (with the convention $b^{-2}(y) = \infty$ for $b(y) = 0$).

The idea of the proof and the details are the same as in the proof of Theorem 4. The only difference is that we now apply Theorem 2 instead of Theorem 1.

REMARK 2. Notice that if (X,F) is a weak solution of (2) on the interval (u,v) and $b(u) = b(v) = 0$ then the process stopped after having reached u or v is a weak solution of (2).
It is interesting to study the question wether a solution of (2) on the interval (u,v) reaches the boundaries u,v or not. We shall investigate this problem in a further note. There we will also extend some results to the case where the functions a,b are time-dependent. An application of Theorem 1 to the problem of the absolute continuity of measures corresponding to (strong) solutions of stochastic differential equations and the Wiener process is given in /3/.

References

/1/ C. Dellacherie: Capacites et processus stochastiques, Springer-Verlag, Berlin Heidelberg New York 1972

/2/ H.J. Engelbert, J. Hess: Stochastic integrals of continuous local martingales I, to appear in: Math. Nachr. 97 (1980)

/3/ H.J. Engelbert, W. Schmidt: A 0-1 law for certain functionals of the Wiener process and some applications, to appear in: C.R. Acad. Bulgare Sci.

/4/ I.I. Gichman, A.V. Skorochod: Stochastic differential equations, Naukovo dumka, Kiev 1968 (in Russian)

/5/ K.Itô, H.P. McKean, Jr.: Diffusion processes and their sample path, Springer-Verlag, Berlin Heidelberg New York 1965

/6/ N. Kazamaki: Changes of time, stochastic integrals, and weak martingales, Z. Wahrscheinlichkeitstheorie verw. Gebiete 22 (1972), 25 - 32

/7/ L.A. Shepp, J.R. Klauder, H. Ezawa: On the divergence of certain functionals of the Wiener process, Ann. Inst. Fourier 24, 2 (1974), 189 - 193

/8/ S. Watanabe: Solution of stochastic differential equations by random time change, Appl. Math. Optim. 2 (1975), 90 - 96

/9/ M.P. Yershov: On stochastic equations, Proc. 2nd Japan-USSR Symp. on Probability Theory, Lecture Notes in Mathematics 330, Springer-Verlag 1973, 527 - 530

FRIEDRICH-SCHILLER-UNIVERSITÄT JENA
SEKTION MATHEMATIK
6900 JENA
UNIVERSITÄTSHOCHHAUS
GERMAN DEMOCRATIC REPUBLIC

POINT PROCESSES AND SYSTEM LIFETIMES

P. Greenwood
University of British Columbia, Vancouver, Canada

1. Introduction and history. A class of problems in lifetime theory can be summarized as follows: Let $T_1,\ldots,T_n$ be non-negative random variables. Let τ be a coherent function of $T_1,\ldots,T_n$, that is

(1) $$\tau = \min_k \max_{i\in A_k} T_i ,$$

where $\{A_k\}$ is a collection of subsets of the indices $1,\ldots,n$. Some hypothesis is made about the joint distribution of $(T_1,\ldots,T_n)$. The question is: what can then be said about the distribution of τ? There is a considerable literature in this direction, and we will review some of it.

In an important early paper, Birnbaum, Esary, and Marshall (1966) studied the question of how the practical problem of "wear-out" of components and systems should be translated into a mathematical property of distribution functions. A natural definition which they consider is: a component or system with lifetime T is wearing out if its failure rate is increasing, or equivalently, if its conditional lifetime is stochastically decreasing on the support of the distribution. More exactly,

$$\int_0^t P(T \in ds)/P(T > s) \text{ is convex, or equivalently}$$

(2) $$P(T > s + t \mid T > t) \text{ is increasing in } t \text{ for}$$

$$t < \inf(s\colon P(T \le s) = 1) .$$

Distributions satisfying (2) they termed IFR for "increasing failure rate". They found the conditions (2) to be unsatisfactory because if a coherent system is formed from components with property (2), the system lifetime need not have this property, even if the components' lifetimes are independent. In their example, T_1, T_2 are independent, exponentially distributed with parameters $\lambda_1 \neq \lambda_2$. Then $\tau_1 = \min(T_1,T_2)$ has property (2), but $\tau_2 = \max(T_1,T_2)$ does not!

Having rejected (2), they considered the condition

(3) $$\frac{1}{t}\int_0^t f(s)/P(T > s)ds \text{ is increasing ,}$$

calling such distributions IFRA, for "increasing failure rate average". They showed that condition (3) has the desired closure property for systems with independent components.

Theorem 1 (Birnbaum, Esary, Marshall). If each independent component of a coherent system is IFRA, then the system is IFRA.

Several attempts were made to extend Theorem 1 to the "multivariate" case, meaning to systems with dependent components. Recently Block and Savits (1980) defined the distribution of $(T_1,\ldots,T_n)$ to be IFRA if for all h continuous, increasing, positive,

(4) $$E(h(T_1,\ldots,T_n)) \le E^{1/\alpha}(h^{\alpha}(T_1/\alpha,\ldots,T_n/\alpha)\,, \quad 0 < \alpha \le 1\,.$$

They showed that (4) is a generalization of (3) and has several closure properties.

<u>Theorem 2</u> (Block and Savits). The family IFRA, (4), is closed under formation of coherent systems, limits in distribution, convolution,... .

At about the same time Ross (1979) gave an elegant proof of the closure of IFRA as defined by (4) in more general terms.

2. <u>Marked Point processes and compensators</u>. An altervative viewpoint on system lifetime theory was suggested by Arjas (1981a,b). Let $(T_1,\ldots,T_n)$ be a lifetime vector and consider the sequence of pairs $(T_{(1)},\xi(T_{(1)}))$, $(T_{(2)},\xi(T_{(2)}))$, ..., where the $T_{(i)}$ are the order statistics of the family $(T_1,\ldots,T_n)$ and $\xi(T_{(i)})$ the associated indices. Let

(5) $$F_t = \sigma((T_{(i)},\xi(T_{(i)}),\ T_{(i)} \le t)\,,$$

the σ-field generated by the lifetimes and indices which have occured before time t. Let

$$Y_t = (N_t,\xi(T_{(1)}),\ldots,\xi(T(N_t)))\,, \qquad N_t = \max(j\colon T_{(j)} \le t)\,.$$

Then Y_t describes a marked point process and F_t is the internal history of Y_t in the terminology of Bremaud and Jacod (1977). An expression for the compensator of a point process and existence and uniqueness theorems have been given by e.g. Liptser and Shiryayev (1978) and also for marked point processes, e.g. Bremaud and Jacod (1977). If $n = 1$, the compensator of Y_t is

(6) $$A(t) = \int_0^{t\wedge T_1} F_1(ds)/1 - F_1(s-)\,,$$

which corresponds to the integrated failure or hazard rate appearing in (2), but stopped at T_1. We note that the denominator here is left- instead of right-continuous.

The correspondence between failure rates and compensators in the simple setting of (6) suggests the following definition, introduced by Arjas (1981b). We simplify the terminology. Let $A_{T,F}$ denote the compensator with respect to σ-fields $F = \{F_t,\ t \ge 0\}$ of the process $1\{T \le t\}$.

<u>Definition</u>. T has <u>F-increasing rate</u> if

(7) $$A_{T,F}(t) \text{ is a.s. convex on } t \le T\,.$$

The class of multivariate distributions such that the marginal distributions have F-increasing rates, where F is defined by (5), has the closure property which Birnbaum, Esary and Marshall wanted.

Theorem 3. If $(T_1,\ldots,T_n)$ is a random vector, $T_i \geq 0$, $P(T_i = T_j) = 0$ each i,j, and each T_i has F-increasing rate, F given by (5), then each coherent system lifetime τ has F-increasing rate.

The proof appears in the next section, with the beginnings of a rudimentary calculus of compensators.

Let us consider Theorem 3 for the example which showed the IFR condition (2), to be an unsatisfactory definition of "wear out". Let T_1,T_2 be independent, exponentially distributed with parameters $\lambda_1 \neq \lambda_2$. Let $\tau = \max(T_1,T_2) = T_{(2)}$. It is easy to check that T_1,T_2 have F-increasing rates and that

$$A_{T_{(2)},F}(t) = \begin{cases} 0, & 0 \leq t \leq T_{(1)} = T_1 \wedge T_2 , \\ \lambda_{\xi(T_{(1)})}(t - T_{(1)}), & T_{(1)} \leq t \leq T_{(2)} , \\ \lambda_{\xi(T_{(1)})}(T_{(2)} - T_{(1)}), & T_{(2)} \leq t , \end{cases}$$

which is convex on $t \leq T_{(2)}$.

Recall that equivalent to the property "increasing failure rate" was the property that the conditional lifetime distribution is stochastically decreasing, as formulated in (2). An F-conditional property of the second sort would be:

Definition. T has F-stochastically decreasing conditional distributions if

(8) $$P(T > s + t | T > t, F_t) \text{ is decreasing in } t ,$$

Arjas showed (1981b) that if each coherent τ of a given multivariate distribution has property (8), then each coherent τ has property (7). But not conversely, as he showed with a simple example. Therefore, one may wish in some circumstances to use (8) as a definition of "closely-watched wear-out" rather than (7). The distribution-valued stochastic process defined by (8) now obtains considerable interest for applications. This process has received essentially no study, compared to its famous relative, the compensator, appearing in (7).

3. Toward a compensator calculus for coherent systems. In the study of coherent systems defined by (1) in terms of a given $(T_1,\ldots,T_n)$, the following stochastic processes arise:

(i) $(T_{(i)},\xi(T_{(i)}))$, the marked point process,

(ii) $N_t = \{\#\text{ of } T_{(i)} \leq t\}$, the counting process associated with (i),

(iii) $(\tau,\xi(\tau))$, a marked point process with just one point, and having as value some subset of $(T_1,\ldots,T_n)$. We call this the marked point associated with a coherent τ.

The families of σ-fields generated by processes (i), (ii), (iii) will be denoted by

(i) $F = \{F_t, t \geq 0\}$

(ii) $\sigma = \{\sigma_t, t \geq 0\}$

(iii) $(\tau) = \{(\tau)_t, t \geq 0\}$.

As above, each compensator appears with two subscripts, the first denoting the process from which it is defined and the second denoting the relevant σ-fields. If a process, say N, is not measurable with respect to a filtration, say (τ), we adopt the convention that $A_{N,(\tau)}$ is the (τ)-compensator of the projection of N with respect to (τ). With this convention, we have at hand nine types of compensators. It is of interest to study some of their interrelations.

As an example, let us compare $A_{N,\sigma}$ and $A_{N,F}$. If we sketch $A_{N,\sigma}$ for a fixed "ω", and then for the same "ω" all possible paths of $A_{N,F}$, we see that $A_{N,F}$ "branches" at each $T_{(i)}$, the various branches corresponding to the possible values of $\xi(T_{(i)})$. By inspection, or by writing out formulas, we can see that if $A_{N,F}$ a.s. convex then $A_{N,\sigma}$ is a.s. convex, whereas the converse statement is false. If the distribution of $(T_1,\ldots,T_n)$ is exchangeable then clearly $A_{N,F} = A_{N,\sigma}$. This observation leads to a one-to-one correspondence between exchangeable distributions of non-negative random vectors and point processes. Accordingly, any partial ordering of exchangeable distributions will correspond to a partial ordering of point processes. In different terms, concepts of positive dependence for exchangeable random vectors correspond to concepts of clustering for point processes.

Let τ be a coherent functional of $(T_1,\ldots,T_n)$. We consider the structure of $A_{\tau,F}$ under the condition $P(T_i = T_j) = 0$, all i,j. Since $1\{\tau \leq t\} = \sum_i 1\{\tau \leq t, \xi_\tau = i\}$ we have

(9) $$A_{\tau,F}(t) = \sum_i A_{(\tau,\xi_\tau=i),F}(t) .$$

We can identify $A_{(\tau,\xi_\tau=i)}$, with a section of the compensator $A_{T_i,F}$ as follows. By definition,

$$\tau = \min_k \max_{i\in A_k} T_i , \quad \text{some } A_k \subset \{1,\ldots,n\} .$$

For each i let σ_i be the coherent functional formed by choosing those A_k containing T_i and removing T_i, i.e.,

$$\sigma_i = \min_{k\in\alpha_i} \max_{j\in V_k^{(i)}} T_j$$

where $V_k^{(i)} = A_k - T_i$, $\alpha_i = \{k\colon T_i \in A_k\}$. We could say that at σ_i, τ is set to occur at T_i. A straightforward computation verifies that

(10) $$A_{(\tau,\xi_\tau=i)}(t) = \int_{\sigma_i}^{\tau\wedge t} A_{T_i,F}(ds) .$$

Proof of Theorem 3. Suppose that $A_{T_i,F}$ are a.s. convex, $i = 1,\dots,n$, $t \le T_i$. Then $\int_{\sigma_i}^{\tau\wedge t} A_{T_i,F}(ds)$ is convex a.s. on $t \le \tau$. From (9) and (10) we have

$$A_{\tau,F}(t) = \sum_i \int_{\sigma_i}^{\tau\wedge t} A_{T_i,F}(ds) ,$$

a sum of a.s. convex functions on $t \le \tau$, and therefore also convex on this interval a.s.

References

Arjas, E. (1981a). A stochastic process approach to multivariate reliability systems: notions based on conditional stochastic order. Math. Op. Res. (to appear)

Arjas, E. (1981b). The failure and hazard processes in multivariate reliability systems. Math. Op. Res. (to appear)

Birnbaum, Z.W. Esary, J.D. and Marshall, A.W. (1966). A stochastic characterization of wear-out for components and systems. Ann. Math. Stat. 37, 816-825.

Block, H.W. and Savits, T.H. (1980). Multivariate increasing hazard rate average distributions. Ann. Prob. 8, 793-801.

Bremaud, P. and Jacod, J. (1977). Processus ponctuels et martingales: resultats récents sur la modélisation et le filtrage. Adv. Appl. Prob. 9, 362-416.

Liptser, R.N. and Shiryayev, A.N. (1978). Statistics of random processes II. Applications, Springer-Verlag New York.

Ross, S.M. (1979). Multivalued state component systems. Ann. Prob. 7, 379-383.

ON WEAK CONVERGENCE OF SEMIMARTINGALES AND POINT PROCESSES

B. Grigelionis, R. Mikulevičius
Institute of Mathematics and Cybernetics
Academy of Sciences of the Lithuanian SSR
University of Vilnius

1. Introduction. As a rule the weak convergence conditions of the sequences of probability measures on topological spaces include assumptions guaranteeing relative compactness of the sequence and some characteristic property for the limiting measure. In the case of weak convergence of semimartingales and point processes it is convenient to express such conditions in the terms of the predictable characteristics and the conditional intensity measures, correspondingly (for terminology see [1]). We shall review some general results of such type in this report. The details of proofs and more complete bibliography can be found in [2] - [3].

2. Characteristic properties of semimartingales. Let $(\Omega,\mathcal{F})$ be a measurable space with increasing right continuous family of σ-algebras, $\mathcal{P}(\mathbb{F})$ be a σ-algebra of $\mathbb{F}$-predictable subsets $R_+ \times \Omega$, $\mathcal{T}(\mathbb{F})$ be a class of $\mathbb{F}$-stopping times, P be a probability measure on $\mathcal{F}$, $R_+ = [0,\infty)$. Denote $\mathcal{M}_{loc}(P,\mathbb{F})$ a class of $(P,\mathbb{F})$ - local martingales, $\mathcal{M}^c_{loc}(P,\mathbb{F})$ a subclass of $\mathcal{M}_{loc}(P,\mathbb{F})$ of the continuous processes, $\mathcal{M}^2_{loc}(P,\mathbb{F})$ a subclass of $\mathcal{M}_{loc}(P,\mathbb{F})$ of the locally square integrable processes and $\mathcal{V}_{loc}(P,\mathbb{F})$ a class of the right continuous $\mathbb{F}$-adapted processes with locally P-integrable variation.

A stochastic process $X = \{X_t = (X^1_t, \ldots, X^m_t), t \geq 0\}$ is said to be a $(P,\mathbb{F})$-semimartingale with the triplet of characteristics (α,β,Π) if it has the following canonical form:

$$X_t = X_0 + \alpha_t + X^c_t + \int_0^t \int_{|x|\leq 1} x\, q(ds,dx) + \int_0^t \int_{|x|>1} x\, p(ds,dx), \quad t \geq 0,$$

where $p(dt,dx)$ is the jump measure of X, $q(dt,dx) = p(dt,dx) - \Pi(dt,dx)$, $\Pi(dt,dx)$ is the $(P,\mathbb{F})$ - dual predictable projection (the conditional intensity measure) of p, $\alpha = (\alpha_1, \ldots, \alpha_m)$ is the $\mathcal{P}(\mathbb{F})$-measurable process,

$$\alpha_j \in \mathcal{V}_{loc}(P,\mathbb{F}), \; j = 1,\ldots,m, \quad \alpha_t - \alpha_{t-} = \int_{|x|\leq 1} x\, \Pi(\{t\}\times dx),$$

$X^c = (X^{c1}, \ldots, X^{cm})$, $X^{cj} \in \mathcal{M}^c_{loc}(P, \mathbb{F})$, $j = 1, \ldots, m$,

$B_t = ||\beta_{jk}(t)||_1^m$, $\beta_{jk}(t) = \langle X^{cj}, X^{ck}\rangle_t$, $t \geq 0$, $j, k = 1, \ldots, m$,

$$\int_0^t \int_E |x|^2 \wedge 1 \Pi(ds, dx) < \infty, \quad t \geq 0; \quad E = R^m \setminus \{0\}.$$

The function

$$\Phi_z(t) = i(z, \alpha_t) - \frac{1}{2}(z, B_t z) + \int_E (e^{i(z,x)} - 1 -$$

$$- \chi_{\{|x| \leq 1\}} i(z,x)) \Pi([0,t] \times dx), \quad z \in R^m, \ t \geq 0,$$

is called the cumulant function of X.

Theorem 1 [2]. The following statements are equivalent:

1) X is a $(P, \mathbb{F})$- semimartingale with the triplet of characteristics (α, B, Π);

2) the measure Π is the $(P, \mathbb{F})$- dual predictable projection of the jump measure of X, for all $z \in R^m$, $(Y, z) \in \mathcal{M}^2_{loc}(P, \mathbb{F})$ and

$$\langle (Y,z)\rangle_t = (z, B_t z) + \int_{|x| \leq 1} (x,z)^2 \Pi([0,t) \times dx) -$$

$$- \sum_{s \leq t} \Big(\int_{|x| \leq 1} (x,z) \Pi(\{s\} \times dx)\Big)^2, \quad t \geq 0,$$

where

$$Y_t = X_t - X_0 - \alpha_t - \int_0^t \int_{|x|>1} x \, p(ds, dx), \quad t \geq 0;$$

3) for all $z \in R^m$ $Y(z) \in \mathcal{M}_{loc}(P, \mathbb{F})$, where

$$Y_t(z) = \exp\{i(X_t - X_0, z)\} - 1 - \int_0^t \exp\{i(X_{s-} -$$

$$- X_0, z)\} \, d\Phi_z(s), \quad t \geq 0;$$

4) for all $f \in C_0^2(R^m)$ [1] $M(f) \in \mathcal{M}_{loc}(P, \mathbb{F})$, where

---

1) $C_0^2(R^m)$ is a class of twice continuously differentiable functions on R^m with the compact support.

$$M_t(f) = f(X_t) - f(X_\cdot) - \sum_{j=1}^{m} \int_0^t D_j f(X_{s-})\, d\,\alpha_j(s) -$$

$$- \frac{1}{2} \sum_{j,k=1}^{m} \int_0^t D^2_{jk} f(X_s)\, d\beta_{jk}(s) - \int_0^t \int_E (f(X_{s-} + x) -$$

$$- f(X_{s-}) - \sum_{j=1}^{m} \chi_{\{|x|\le 1\}} D_j f(X_{s-})\, x_j)\, \Pi(ds,dx),\ t \ge 0.$$

Exploiting these characteristic properties of semimartingales we can obtain several forms of limit theorems for semimartingales.

3. Weak convergence of semimartingales. Let $X_n = \{X_n(t),\ t \ge 0\}$ be a m-dimensional semimartingale on the probability space $(\Omega_n, \mathcal{F}_n, P_n)$ having an initial distribution $\nu^{(n)}$, a triplet $(\alpha^{(n)}, B^{(n)}, \Pi^{(n)})$ of the predictable characteristics with respect to P_n and a filtration $\mathbb{F}_n = \{\mathcal{F}_{nt},\ t \ge 0\}$ of σ-algebras and a cumulant function $\Phi_z^{(n)}(t)$. Denote $\hat{\Omega} = D_{[0,\infty)}(R^m)$ with the $\mathcal{J}_1$-topology of Skorokhod and the standard filtration $\hat{\mathbb{F}} = \{\hat{\mathcal{F}}_t,\ t \ge 0\}$ of σ-algebras, $\hat{\mathcal{F}} = \bigvee_{t\ge 0} \hat{\mathcal{F}}_t$, $\hat{X}_t(\hat{\omega}) = \hat{\omega}(t),\ t \ge 0$, $\hat{P}_n = P_n \circ X_n^{-1},\ n \ge 1$, and the measure $\hat{P}$ on $\hat{\mathcal{F}}$ such that $\hat{X}$ is a $(\hat{P}, \hat{\mathbb{F}})$ - semimartingale having an initial distribution $\hat{\nu}$, a triplet $(\hat{\alpha}, \hat{B}, \hat{\Pi})$ of the predictable characteristics, uniquely defining $\hat{P}$, and a cumulant function $\hat{\Phi}_z(t, \hat{\omega})$. Let $\mathcal{S}(T)$ be a class of sequences $\{S_n,\ n \ge 1\}$ such that $S_n \in \mathcal{T}(\mathbb{F}_n),\ n \ge 1$ and

$$\lim_{n\to\infty} P_n \{S_n < T\} = 0,\ T > 0,$$

and $\mathcal{K}$ be a class of increasing functions $G: R_+ \to R_+$, such that $G(t)/t \to \infty$ as $t \to \infty$.

Theorem 2 (cf. [2]). Assume that:

1) the sequence $\{\hat{P}_n,\ n \ge 1\}$ is tight;

2) $\nu^{(n)} \Rightarrow \hat{\nu}$;

3) for each $T > 0$ and $z \in R^m$ there exist $\{S_n\} \in \mathcal{S}(T)$ and $G \in \mathcal{K}$ such that

$$\sup_{n\ge 1} E_n [G(|\Phi_z^{(n)}|(S_n))] < \infty,$$

where $|\Phi|(t)$ denotes the variation of $\Phi(t)$ on $[0,t]$;

4) for each $z \in R^m$, $\varepsilon > 0$ and $t \in Q$ [1/]

$$\lim_{n\to\infty} P_n \{|\Phi_z^{(n)}(t) - \hat{\Phi}_z (t,X_n)| > \varepsilon\} = 0;$$

5) $\hat{\Pi}(\{t\} \times E) \equiv 0$ and for all $z \in R^m$ and $t \in Q$ $\hat{\Phi}_z(t,.)$ is $\mathcal{J}_1$-continuous.

Then $\hat{P}_n \Rightarrow \hat{P}$.

Remark 1. Applying [4] and Theorem 1 it is easy to check that $\{\hat{P}_n, n \geq 1\}$ is tight if for each $T > 0$, $\varepsilon > 0$, $z \in R^m$, $\{\delta_n, n \geq 1\}$, $\{T_n, n \geq 1\}$ such that $\delta_n \downarrow 0$ as $n\to\infty$, $T_n \in \mathcal{T}(F_n)$, $T_n \leq T$ and T_n takes only finite number of values, there exists $\{S_n\} \in \mathcal{S}(T)$ such that

$$E_n \ |\Phi_z^{(n)}|(S_n) < \infty, \ n \geq 1,$$

$$\lim_{n\to\infty} E_n[|\Phi_z^{(n)}| \ (T_n + \delta_n)\wedge S_n) - |\Phi_z^{(n)}|(T_n\wedge S_n)] = 0$$

and

$$\lim_{L\to\infty} \overline{\lim_{n\to\infty}} \ P_n \ \{\Pi^{(n)}([0,T] \times \{|x| > L\}) \geq \varepsilon\}.$$

Some other forms of the conditions for the tightness and the weak convergence of $\{\hat{P}_n, n \geq 1\}$ can also be found (see, e.g., [2], [4]-[8]). Let $\mathcal{U}$ be a class of nonnegative continuous functions on E with the compact support,

$$\Gamma_z^{(n)}(t) = (z, B_t^{(n)} z) + \int_0^t \int_{|x|\leq 1} (x,z)^2 \ \Pi^{(n)} (ds,dx) -$$

$$- \sum_{s\leq t} \Big(\int_{|x|\leq 1} (x, z) \ \Pi^{(n)} (\{s\} \ xdx)\Big)^2, \ n \geq 1,$$

$$\hat{\Gamma}_z (t,\hat{\omega}) = (z, \hat{B}_t(\hat{\omega})z) + \int_0^t \int_{|x|\leq 1} (x,z)^2 \ \hat{\Pi}(\hat{\omega},ds,dx) -$$

$$- \sum_{s\leq t} \Big(\int_{|x|\leq 1} (x,z) \ \Pi(\hat{\omega}, \{s\} \ xdx)\Big)^2, \ t \geq 0, \ z \in R^m, \ \hat{\omega} \in \hat{\Omega}.$$

[1/] Q denotes some denumerable dense subset of R_+.

<u>Theorem 3 [2]</u>. Assume that:

1) the sequence $\{\hat{P}_n, n \geq 1\}$ is tight;

2) $\nu^{(n)} \Rightarrow \hat{\nu}$;

3) for each $\varepsilon > 0$, $z \in R^m$, $\varphi \in \mathcal{U}$ and $t \in Q$

$$\lim_{n\to\infty} P_n \{|\Gamma_z^{(n)}(t) - \hat{\Gamma}_z(t, X_n)| < \varepsilon\} = 0,$$

$$\lim_{n\to\infty} P_n \{|\alpha_t^{(n)} - \hat{\alpha}_t(X_n)| > \varepsilon\} = 0,$$

$$\lim_{n\to\infty} P_n \{|\int_E \varphi(x)\, \Pi^{(n)}([0,t]\times dx) - \int_E \varphi(x)\, \hat{\Pi}(X_n, [0,t]\times dx)| > \varepsilon\} = 0;$$

4) for each $T > 0$, $z \in R^m$, $\varphi \in \mathcal{U}$ there exist $\{S_n\} \in \mathcal{S}(T)$ and $G_1, G_2 \in \mathcal{K}$ such that

$$\sup_{n\geq 1} E_n\, G_1(\Gamma_z^{(n)}(S_n)) < \infty, \sup_{n\geq 1} E_n\, G_2(\int_E \varphi(x)\, \Pi^{(n)}([0, S_n]\times dx)) < \infty;$$

5) for each $z \in R^m$, $\varphi \in \mathcal{U}$ and $t \in Q$ the functions $\hat{\alpha}(t,.)$, $\hat{\Gamma}_z(t,.)$ and $\int_E \varphi(x)\, \hat{\Pi}(.,[0,t]\times dx)$ are $\mathcal{J}_1$-continuous and $\hat{\Pi}([0,t]\times\{y: |y| = 1\}) \equiv 0$ $\hat{P}$-a.e. .

Then $\hat{P}_n \Rightarrow \hat{P}$.

Note that if X is the process with the homogeneous independent increments, then the cumulant function is

$$\Phi_z(t) = t \ln E \exp\{i(z, X_1 - X_o)\}$$

and it uniqely defines the corresponding measure with given initial distribution (see [9]).

If we consider a sequence of semimartingales

$$X_n(t) = \sum_{t_{nk} \leq t} X_{nk},\ t \geq 0,\ n \geq 1,$$

where $0 = t_{no} < t_{n1} < \ldots,\ t_{n_k} \uparrow \infty$ as $k \to \infty$, then it is easy to cehck that

$$\Phi_z^{(n)}(t) = \sum_{t_{nk} \leq t} \int_E (e^{i(z,x)} - 1)\, P_{nk}(dx),$$

where $\qquad P_{n_k}(\Gamma) = P_n\{X_{n_k} \in \Gamma \mid \mathcal{F}_{n,\, t_{n,k-1}}\}.$

4. <u>Weak convergence of point processes.</u> Let now (E,ρ) be a separable metric locally compact space, $\tilde{E} = R_+ \times E$, $\tilde{\mathcal{E}} = \mathcal{B}(R_+) \otimes \mathcal{B}(E)$, $\tilde{\Omega}$ be a space of the nonnegative integer-valued Radon measures $\tilde{\omega}(dt,dx)$ such that $\tilde{\omega}(\{0\} \times E) = 0$, $\tilde{\omega}(\{t\} \times E) \le 1$ for all $t \ge 0$, with the topology of vague convergence, the standard filtration $\mathbb{F} = \{\tilde{\mathcal{F}}_t,\ t \ge 0\}$ of σ-algebras, $\tilde{\mathcal{F}} = \bigvee_{t\ge 0} \tilde{\mathcal{F}}_t$, $\tilde{p}(\tilde{\omega},dt,dx) = \tilde{\omega}(dt,dx)$. Let $p_n = \{p_n(\tilde{B}),\ B \in \tilde{\mathcal{E}}\}$ be a point process on $(\Omega_n, \mathcal{F}_n, P_n)$ with values in $\tilde{\Omega}$ having the $(P_n, \mathbb{F}_n)$-dual predictable projection (the conditional intensity measure) $\Pi^{(n)}$, $\tilde{P}^{(n)} = P_n \circ p_n^{-1}$ and $\tilde{P}$ be a measure on $\tilde{\mathcal{F}}$ such that $\tilde{p}$ has the $(\tilde{P}, \tilde{\mathbb{F}})$-dual predictable projection $\tilde{\Pi}$, uniquely defining $\tilde{p}$.

<u>Theorem 4</u>. Assume that:

1) the sequence $\{\tilde{P}_n,\ n \ge 1\}$ is relatively compact;

2) for all nonnegative $f \in C_0(E)$, $\varepsilon > 0$, $t \in Q$

$$\lim_{n\to\infty} P_n\{|\Pi_t^{(n)}(f) - \tilde{\Pi}_t(p_n, f)| > \varepsilon\} = 0,$$

where

$$\Pi_t^{(n)}(f) = \int_0^t \int_E f(x)\, \Pi^{(n)}(ds,dx),\quad \tilde{\Pi}_t(\tilde{\omega},f) = \int_0^t \int_E f(x)\, \tilde{\Pi}(\tilde{\omega}, ds,dx);$$

3) for all $T > 0$ and $f \in C_0(E)$ there exist $G \in \mathcal{K}$ and sequence $\{S_n\} \in \mathcal{S}(T)$ such that

$$\sup_{n\ge 1} E_n\,[G(\Pi^{(n)}_{S_n \wedge T}(|f|))] < \infty;$$

4) for all $t\in Q$ and $f \in C_0(E)$ $\tilde{\Pi}_t(\tilde{\omega}, f)$ is vaguely continuous in $\tilde{\omega}$.

Then $\tilde{P}_n \Rightarrow \tilde{P}$.

Denote

$$p_n^K = \{P_n^K(t) = p_n([0,t]\times K),\ t \ge 0\},$$

$$\tilde{P}_n^K = P_n \circ (p_n^K)^{-1},\ n \ge 1,\ K \in \mathcal{B}(E).$$

The following criterion is true.

Theorem 5. The sequence $\{\tilde{P}_n, n \geq 1\}$ is relatively compact iff for all nonnegative $f \in C_0(E)$, $k \geq 1$

$$\lim_{L\to\infty} \sup_{n\geq 1} P_n \{\Pi_k^{(n)}(f) > L\} = 0$$

and the sequence $\{\tilde{P}_n^K, n \geq 1\}$ is relatively compact on $D_{[0,\infty)}(R)$ with the $\mathcal{J}_1$-topology of Skorokhod for all relatively compact subsets $K \subset E$.

Remark 2. According to [5] the sequence $\{\tilde{P}_n^K, n \geq 1\}$ is relatively compact on $D_{[0,\infty)}(R)$ if there exist a sequence of $\mathcal{P}(\mathbf{F}_n)$-measurable increasing processes $G_n^K(t)$, $t \geq 0$, and a nonrandom increasing function $G_\infty^K(t)$, $t \geq 0$, such that $G_n^K(t) - \Pi_t^{(n)}(X_K)$, $t \geq 0$, is also an increasing process and for all $\varepsilon > 0$, $t \in Q$

$$\lim_{n\to\infty} P_n \{|G_n^K(t) - G_\infty^K(t)| > \varepsilon\} = 0$$

and

$$\lim_{n\to\infty} P_n\{| \sum_{u\leq t} [(\Delta G_n^K(u))^2 - (\Delta G_\infty^K(u))^2] > \varepsilon\} = 0,$$

where $\quad \Delta f(t) = f(t) - f(t-)$.

References

[1] J. Jacod, Calcul stochastique et problemes de martingales. - Lecture Notes in Math., 714, Springer, 1979.

[2] B. Grigelionis, R. Mikulevičius, On weak convergence of semimartingales. - Lietuvos matem. rink., 1981, vol. XXI, No 1.

[3] B. Grigelionis, R. Mikulevičius, On weak convergence of random point processes. - Lietuvos matem. rink., 1981, vol. XXI, No 4.

[4] D. Aldous, Stopping times and tightness. - Ann. Probab., 1978, vol. 6, p. 335-340.

[5] J. Jacod, J. Mémin, Un nouveau critere de compacité relative pour une suite de processes. - Sém. de Probab. Rennes, 1979.

[6] J. Jacod, J. Mémin, Sur la convergence des semimartingales vers un processus a accroissements independents. - Sém. Probab. Strasbourg XIV, Lecture Notes in Math., 784, Springer, 1979.

[7] R. Rebolledo, La methode des martingales applique a l'etude de la convergence en loi de processus. - Bull. de la Société Mathematique de France, Mém. No. 62, 1979.

[8] M. Metivier, Une condition suffisante de compasite faible pour une suite de procqssus (preprint), 1980.

[9] B. Grigelionis, On martingale characterization of stochastic processes with independent increments. - Lietuvos matem. rink., 1977, vol. XVII, No 1, p. 75-86.

B. Grigelionis

Institute of Mathematics and Cybernetics
Academy of Sciences of the Lithuanian SSR,
232 600 Vilnius 54, K. Pozelos str.
U.S.S.R.

R. Mikulevicius

Institute of Mathematics and Cybernetics
Academy of Sciences of the Lithuanian SSR,
232 600 Vilnius 54, K. Pozelos str.
U.S.S.R.

ITO FORMULA IN BANACH SPACES

I. Gyöngy
Eötvös Loránd University Budapest
Department of Algebra and Number Theory

N.V. Krylov
Lomonosov University Moscow
Department of Probability Theory

1. Introduction

The type of Ito formula we are concerned within this paper arised in the course of dealing with stochastical partial differential equations /SPDE/. In many cases SPREs can be considered as stochastical differential equations /SDE/ in infinite dimensional Banach spaces. Usually the well-known Ito formula is of course an essential tool in the study of SDEs, but if the coefficients of the SDE we consider are unbounded operators, which occurs at certain SPREs, then a new type of Ito formula is necessary. Following Pardoux in using the Lions' scheme, the situation is as follows.

Let V be a separable Banach space, which is continuously embedded into a separable Hilbert space H such that V is dense in H. The space H is identified with its dual space H^* /by the scalar product in H/, consequently we have

$$V \subset H \equiv H^* \subset V^* \ ,$$

where V^* is the dual space of V and the embedding $H^* \subset V^*$ is the adjoint embedding of that of $V \subset H$. We are given a V^*-valued local semimartingale of the form

$$y(t) := \int_{]0,t]} v^*(u)dA(u) + h(t) \ ,$$

where $v^*(t)$ is a V^*-valued process, $A(t)$ is a real valued increasing process and $h(t)$ is an H-valued locally square integrable martingale on a fixed probability space $(\Omega, \mathcal{F}, P)$ endowed with an increasing family of sub-σ-fields of $\mathcal{F}$. Moreover there is given a V-valued process $v(t)$ such that $dP \times dA(t)$-almost everywhere $v(t) = y(t)$ in V^*. One wants to have a kind of Ito formula for $y^2(t)$, where y^2 denotes the scalar product of y by itself in H. The first question which arises here is the following: since $y(t)$ takes its values in V^* and generally H is a proper subset of V^*, is it possible to consider $y(t)$ as an H-valued process. It will be a consequence of the main theorem of this paper that under natural measurability conditions on the processes $v(t)$, $v^*(t)$, $A(t)$ and under the assumption that $|v(t)|_V$, $|v^*(t)|_{V^*}$ and $|v(t)|_V \, |v^*(t)|_{V^*}$ are almost surely locally integrable /with respect to $dA(t)$/ $y(t)$ is — up to indistinguishability — an H-valued adapted cadlag process and for $y^2(t)$ the Ito formula is valid. A detailed discussion of Ito formula in Hilbert spaces can be found in Metivier [3].

The possibility and the importance of the Ito formula for $y^2(t)$ was firstly shown by E. Pardoux /see [4] and [5]/ in the case when $A(t) = t$ and $h(t)$ is continuous. The Ito formula in [4] is proved using the theory of Ito equations in Banach spaces developed by E. Pardoux in the mentioned paper, and it is achieved under some assumptions connected with that theory. A self-contained proof under only natural measurability and some integrability conditions is done in Krylov-Rozovskii [2] for the same case of continuous $h(t)$ and $A(t) = t$. In this paper we present some general results for the case when $A(t)$ is an increasing adapted cadlag process and $h(t)$ is an H-valued locally square integrabel cadlag martingale. We formulate our results only on the Ito formula for $y^2(t)$ because this is the most important case if one uses Ito formula in the study of SDEs. We note that our results can be generalized for other functions of $y(t)$ as well. The detailed proofs of these results can be found in the forthcoming paper [1].

2. Assumptions and basic theorems

Let V be a separable Banach space and V^* its dual space. Suppose that there exists a separable Hilbert space H and a bounded linear operator $\lambda : V \to H$ such that λV is dense in H. We denote by uz the scalar product of $u, z \in H$ and use the same notation for the duality product between V and V^* if one of the elements u, z belongs to V and the other to V^*. For an element u from a Banach space we denote by $|u|$ the norm of u. We fix a complete probability space $(\Omega, \mathcal{F}, P)$ and an increasing family of σ-fields $(\mathcal{F}_t)_{t\geq 0}$ /$\mathcal{F}_t \subset \mathcal{F}$/ with the usual conditions: $\mathcal{F}_t = \bigcap_{\rho>t} \mathcal{F}_\rho$, $\mathcal{F}_0$ contains all the P-null sets of $\mathcal{F}$. Let h be an H-valued locally square integrable /strongly/ cadlag martingale, $A(t)$ a real-valued increasing adapted cadlag process starting from zero, $v(t)$ a

V-valued process such that $vv^*(t)$ is progressively measurable for every $v\in V$. Suppose that $|v(t)|$, $|v^*(t)|$ and $|v(t)|\,|v^*(t)|$ are almost surely locally integrable with respect to $dA(t)$.

Now we formulate our main theorem.

<u>Theorem 1.</u> Let τ be a stopping time. Suppose that for every $v\in V$ for $dP\times dA(t)$-almost all $(\omega,t)\in]0,\tau[$

$$\Lambda v\Lambda v(t) = \int_{]0,t]} vv^*(u)dA(u) + \Lambda vh(t)\ .$$

Then there eists a subset $\tilde{\Omega}\subset\Omega$ with $P(\tilde{\Omega}) = 1$ and an H-valued adapted cadlag process $h(t)$ such that

$$h(t) = \Lambda v(t)$$

for $dP\times dA(t)$-almost all $(\omega, t)\in]0,\tau[$, moreover for every $\omega\in\tilde{\Omega}$ and $t<\tau(\omega)$ we have

$$\Lambda vh(t) = \int_{]0,t]} vv^*(u)dA(u) + \Lambda vh(t)$$

for every $v\in V$, and

$$\tilde{h}^2(t) = h^2(0) + 2\int_{]0,t]} v(u)v^*(u)dA(u) + 2\int_{]0,t]} h(u-)dh(u) -$$

$$- \int_{]0,t]} |\Lambda^{*-1}v^*(u)|^2\Lambda A(u)dA(u) + [h]_t,$$

where we set $|\Lambda^{*-1}v^*(u)| := \infty$ if $v^*(u)\notin\Lambda^*H^*$.

Let us consider now the special case when $V\subset H$ and Λ is the identy on V. We suppose that V is dense in H and that with a constant K, $|v|_H \le K|v|_V$ for every $v\in V$, and moreover suppose that H is a dense subset of some separable Banach space V' and $|\varphi\psi| \le |\varphi|_V|\psi|_{V'}$ for every $\varphi\in V$ and $\psi\in H$. Then for every $\psi\in V'$ we can uniquely define a continuous linear functional on V by $\varphi\psi := \lim_{n\to\infty}\varphi\psi_n$, where $\varphi\in V$, $\psi_n\in H$ and $|\psi-\psi_n|_{V'} \to 0$. Suppose that the bounded linear operator $T : V' \to V^*$ we get in this way is one-to-one. Let $v'(t)$ be a V'-valued progressively measurable process such that $|v'(t)|$ is almost surely locally integrable with respect to $dA(t)$. Then — as it is easy to see — there exists $\Omega'\subset\Omega$

with $P(\Omega') = 1$ such that for every $\omega\in\Omega'$ the Bochner-integral $\int_{]0,t]} v'(u)dA(u)$ for all $t \geq 0$ exists. Now we formulate our next result.

Theorem 2. Let $\tilde{h}(t) = \int_{]0,t]} v'(u)dA(u) + h(t)$ for $\omega\in\Omega'$ and $h(t) = 0\in V'$ elsewhere. Suppose that $\tilde{h}(t) = v(t)$ for $dP\times dA(t)$-almost all (t,ω). Then there exist $\Omega''\subset\Omega'$ with $P(\Omega'') = 1$ such that $\chi_{\Omega''}h(t)$ is an H-valued adapted cadlag process and for $\omega\in\Omega''$ and $t \geq 0$

$$\tilde{h}^2(t) = h^2(0) + 2\int_{]0,t]} \tilde{h}(u)v'(u)dA(u) + 2\int_{]0,t]} \tilde{h}(u-)dh(u) -$$

$$- \int_{]0,t]} |v'(u)|_H^2 \Delta A(u)dA(u) + [h]_t,$$

where $|v'(u)|_H := \infty$ if $v'(u)\notin H$, $\tilde{h}(u)v'(u) := \infty$ if $\tilde{h}(u)\notin V$.

This theorem can be obtained from Theorem 1, by considering the V^*-valued process $Tv'(t)$ instead of $v^*(t)$ and noting that $Tv'(t)\in\Lambda^*H^*$ if and only if $v'\in H$.

This theorem is obtained under minimal assumptions on measur ability properties of $A(t)$, $v(t)$, $v'(t)$.

Using the same assumptions, but instead of the progressively measurability of $A(t)$ and $v'(t)$ we assume that they are predictable processes, then from Theorem 2 we get the following.

Corollary. There exists a real valued local martingale $m(t)$ such that for every $\omega\in\Omega''$ and $t \geq 0$

$$\tilde{h}^2(t) = h^2(0) + 2\int_{]0,t]} \tilde{h}(u-)v'(u)dA(u) + 2\int_{]0,t]} \tilde{h}(u-)\, dh(u) +$$

$$+ \int_{]0,t]} |v'(u)|_H^2 \Delta A(u)dA(u) + [h]_t + m(t).$$

<u>References</u>

[1] Gyöngy, I. and Krylov, N.V.: On stochastic equations with respect to semimartingales II., Ito formula in Banach spaces, to appear

[2] Krylov, N.V. and Rozovskii, V.L.: Ito equations in Banach spaces, Itogi nauki, Teor, verojatn. /VINITI AN USSR/, 71-146 /1979/.

[3] Metivier, M.: Reele und Vectorwertige Quasimartingale und die Theorie der Stochastischen Integration, Lect. Notes Math. 607 /1977/, Springer-Verlag.

[4] Pardoux, E.: Thèse, Univ.de Paris-Sud, 1975.

[5] Pardoux, E.: Stochastic Partial Differential Equations and Filtering of Diffusion Processes, Stochastics, 3, 127-167 /1979/.

I. Gyöngy

Eötvös Loránd University Budapest
Department of Algebra and Number Theory
1088 Budapest, Muzeum krt. 6-8.

N.V. Krylov

Lomonosov University
Moscow
Department of Probability Theory

GENERAL THEOREMS OF FILTERING WITH POINT PROCESS OBSERVATIONS

Dimitar I. Hadžiev
Institute of Mathematics
Bulgarian Academy of Sciences,
Sofia, BULGARIA

1. Introduction. Let $(\Omega, \mathcal{F}_\infty, P)$ be a complete probability space equipped with a nondecreasing and right-continuous family of sub-σ-algebras $\mathcal{F} = (\mathcal{F}_t)$, $t \in R_+ = [0,\infty)$, where $\mathcal{F}_0$ is completed with $\mathbb{P}$-null sets.

We discuss an arbitrary $\mathcal{F}$-adapted cadlag semi-martingale $Y = (Y_t)$, $t \in R_+$, with natural representation

$$Y = Y_0 + B + M, \tag{1}$$

where $\mathbb{E}|Y_0|<\infty$, $B = (B_t)$ is $\mathcal{F}$-predictable process with integrable on $(0, t]$ variation ($B_0 = 0$, $\mathbb{E} \int_{(0,t]} |dB_s|<\infty$, $t \in R_+$) and $M = (M_t)$ is a cadlag martingale with respect to $\mathcal{F}(M_0 = 0$, $\mathbb{E}(M_{t+s}/\mathcal{F}_t) = M_t$, $\mathbb{E}|M_t|<\infty$, $s, t \in R_+)$.

Suppose that one cannot observe directly the process Y, but one is able to observe successively at $\mathcal{F}$-stopping time T_n some random variable $X_n = X_n(\omega)$, $n \geq 1$, taking values from a measurable space $(E,\mathcal{E})$. We assume that

(i) $T_n \leq T_{n+1}$, $n \geq 1$;

(ii) X_n is $\mathcal{F}_{T_n}$-measurable;

(iii) $T_n < T_{n+1}$ on $\{\omega \in \Omega: T_n(\omega)<\infty\}$, $n \geq 0$, $T_0 = 0$.

Then (T_n, X_n), $n \geq 1$, form a (multivariate) point process as defined by Jacod [2].

Let $\mu(\omega, dt, dx)$ be the corresponding random measure

$$\mu(\omega, dt, dx) = \sum_{n\geq 1} 1_{\{T_n<\infty\}} \cdot \varepsilon_{(T_n, X_n)}(dt, dx) \tag{2}$$

where $\varepsilon_Z(\cdot)$ is the Dirac measure located at Z, and $1_{\{\cdot\}}$ is the indicator of $\{\cdot\}$. We introduce the (completed w.r.t. P) family of sub-σ-algebras $\mathcal{F}^\mu = \mathcal{F}^\mu_t$ with

$$\mathcal{F}^\mu_t = \sigma\{\mu(\omega, (0,s]\times C), s \leq t, C \in \mathcal{E}\} \subseteq \mathcal{F}_t, \tag{3}$$

which is also right-continuous and nondecreasing. The family $\mathcal{F}^{\mu}$ will be interpreted as a flow of observations.

The problem of optimal filtering presented in [4], is to give an integral representation of the optimal filter $\Pi = (\Pi_t)$ which is the unique (up to indistinguishability, [1]) cadlag modification of the process $E(Y_t/\mathcal{F}_t^{\mu})$, $t \in R_+$. We give in this paper two general theorems describing the structure of the process Π.

2. Notations and preliminaries. The optimal filter Π will be described in the terms of: the dual $\mathcal{F}^{\mu}$-predictable projection $\bar{B}$ of the "drift" B in (1), the $\mathcal{F}^{\mu}$-predictable projection $\bar{\nu}(\omega, dt, dx)$ of the measure μ from (2), and the predictable expectation $\langle Y/\mu \rangle$ of the process Y with respect to the random measure μ.

The dual $\mathcal{F}^{\mu}$-predictable projection $\bar{B}$ (in the sense of Dellacherie [1]) is the unique $\mathcal{F}^{\mu}$-predictable process with integrable on (0, t] variation (and $\bar{B}_0 = 0$) such that $\mathbb{E} \int_{(0,T]} |d\bar{B}_s| = \mathbb{E} \int_{(0,T]} |dB_s|$ for every $\mathcal{F}^{\mu}$-stopping time $T = T(\omega)$.

Let $\mathcal{P}_0(\mathcal{F}^{\mu})$ be the σ-algebra of $\mathcal{F}^{\mu}$-predictable subsets of $\Omega \times R_+$ and $\bar{\mathcal{P}} = \mathcal{P}_0(\mathcal{F}^{\mu}) \otimes \mathcal{E}$. Let $(E, \mathcal{E})$ be an absolutely measurable subset of a compact metric space with Borel σ-algebra. Jacod has proved that there exists a (unique) random measure $\bar{\nu}(\omega, dt, dx)$ such that

a) for every $\bar{\mathcal{P}}$-measurable variable $U = U(\omega, t, x)$ the process $U * \bar{\nu}$ with

$$U * \bar{\nu}_t(\omega) = \int_{(0,t] \times E} U(\omega, s, x)\, \bar{\nu}(\omega, ds, dx) \tag{4}$$

is $\mathcal{F}^{\mu}$-predictable;

b) the measure M_{μ}, where

$$M_{\mu}(U) = \mathbb{E}(U * \mu_{\infty}) \tag{5}$$

coincides with the measure $M_{\bar{\nu}}$ on $\bar{\mathcal{P}}$, and

c) $\bar{\nu}(\omega, \{t\} \times E) \le 1$, $t \in R_+$, and $\bar{\nu}(\omega, [T_{\infty}, \infty) \times E) = 0$,

where $T_{\infty}(\omega) = \uparrow \lim T_n(\omega)$.

This $\bar{\nu}$ is called ([2]) the $\mathcal{F}^{\mu}$-predictable projection of the random measure μ from (2). We denote as well

$$\bar{a}_t(\omega) = \bar{\nu}(\omega, \{t\} \times E), \quad \hat{U}_t(\nu) = \int_E U(\omega, t, x)\bar{\nu}(\omega, \{t\}, dx). \tag{6}$$

Let $Y\cdot\mu(\omega, dt, dx) = Y_t(\omega)\mu(\omega, dt, dx)$. The Radon-Nikodym derivative of the restriction of $M_{Y.\mu}$ on Φ with respect to the restriction of M_μ on Φ permits a Φ-measurable version $\langle Y/\mu\rangle$ such that

(7) $$\widehat{\langle Y/\mu\rangle} = \Pi_- + \Delta\bar{B} \quad \text{on} \quad \{(\omega,t) : \bar{a}_t(\omega) = 1\},$$

where $\Pi_- = (\Pi_{t-})$, $\Pi_{t-} = \lim_{s\uparrow t} \Pi_s$, and $\Delta\bar{B} = (\Delta\bar{B}_t)$, $\Delta\bar{B}_t = \bar{B}_t - \bar{B}_{t-}$, $t\in R_+$ (see [4]). This $\langle Y/\mu\rangle$ is called the predictable expectation of the process Y with respect to the random measure μ.

We remind ([2]), that $\bar{\nu}$ is easily expressed in terms of the conditional distributions of the observations (T_n, X_n), $n \geq 1$. The term $\bar{B}$ is connected with $\mathfrak{F}^\mu$ and the drift B of Y only. Generally speaking $\bar{B}$ is usually calculated without principle difficulties. That is the reason why μ, $\bar{\nu}$ and $\bar{B}$ will be interpreted as "previously known" characteristics. On the contrary, $\langle Y/\mu\rangle$ includes the whole of the non-observable process Y and therefore must be considered as a solution of a special estimation problem.

3. Representation theorem. The objects μ, $\bar{\nu}$, $\bar{B}$ and $\langle Y/\mu\rangle$ characterize completely the optimal filter Π.

Theorem 1. On $[0,T_\infty[= \{(\omega,t) : 0 \leq t < T_\infty(\omega)\}$ the process Π permits the representation

(8) $$\Pi = \Pi_0 + \bar{B} + Z*(\mu-\bar{\nu}),$$

where

(9) $$\begin{cases} Z = V + \hat{V}\,(1-\bar{a})^\oplus \\ V = \langle Y|\mu\rangle - \Pi_- - \Delta\bar{B} \end{cases}$$

and $b^\oplus = b^{-1}$, if $b \neq 0$, and $b^\oplus = 0$, if $b = 0$.

This theorem has been proved in [4]. It tells that Π is a semimartingale whose martingale part can be explicitely expressed in terms of $\langle Y/\mu\rangle$ and the "previously known" characteristics μ, $\bar{\nu}$, $\bar{B}$.

4. The predictable expectation. In this section we state that the problem of calculating the predictable expectation $\langle Y/\mu\rangle$ can be reduced to a discrete filtering problem.

Let x_o, x_∞ be two different points which are isolated for the metric space $(E,\mathcal{E})$. We define $\eta = (N,\xi)$, where

$$(10)\quad \xi_t(\omega) = \begin{cases} X_n(\omega), & \text{if } T_n \le t < T_{n+1},\ X_o(\omega) = x_o, \\ x_\infty, & \text{if } t \ge T_\infty, \end{cases}$$

and

$$(11)\qquad N_t(\omega) = \mu(\omega, (0,t] \times E),\ t \in R_+.$$

The precise result follows.

Theorem 2. Let

(j) $\quad \mathbb{E}(|\theta_n|)<\infty$, where $\theta_n = Y_{T_n} 1_{\{T_n<\infty\}}$, $n \ge 1$,

and

(jj) for every $\omega \in \Omega$, $t \in R_+$, there exists $\omega' \in \Omega$ such that

$$\eta_s(\omega') = \eta_{t\wedge s}(\omega),\ s \in R_+.$$

(This means that Ω is rich enough)

Then 1) there exist Borel functions $g_n = g_n(t_1, x_1,\ldots,t_n, x_n)$, $n \ge 1$, for which

$$(12)\qquad g_n(T_1, X_1,\ldots,T_n, X_n) = \mathbb{E}(\theta_n/\mathcal{F}^{\mu}_{T_n}),$$

and

2) on $[\![0,T_\infty[\![$ we have

$$(13)\qquad \langle Y/\mu\rangle = \sum_{n\ge 0} g_{n+1}(T_1, X_1,\ldots,T_n, X_n, t, x)\, 1_{\{T_n < t \le T_{n+1}\}}.$$

The proof of this theorem is going to be published in [6]. We shall only note here that the functions g_n, $n \ge 1$, are just regular versions of the conditional expectations

$$\mathbb{E}(Y_{T_n} 1_{\{T_n<\infty\}}/ T_1, X_1,\ldots T_n,X_n),\ n \ge 1.$$

So in general the evaluation of $\langle Y/\mu\rangle$ depends on the effective filtering problem solution for the sequence θ_n, $n \ge 1$, from the observations of the sequence (T_n, X_n), $n \ge 1$.

Remark. The particular case E = {1} is just the case of observing the counting process N from (11). It has already been treated in [5] where an example is included too. Many other examples of good computation one can find in [3], Ch. 19. Another type of examples which illustrate general Theorems 1 and 2 will be published in a forthcoming paper.

REFERENCES

[1] Dellacherie,C.: Capacites et processus stochastiques, Springer-Verlag, Berlin (1972).

[2] Jacod,J.: Multivariate point processes: predictable projection, Radon-Nikodym derivatives representation of martingales. Z. Wahr.verw.Geb. 31, 235-253 (1975).

[3] Liptser,R.S., Shiryayev,A.N.: Statistics of random processes II, Applications. Springer-Verlag, Berlin (1978).

[4] Hadžiev,D.I.: On the filtering of semi-martingales in case of observations of point processes. Theory Prob.Appl. 23, 175-184 (1978).

[5] Hadžiev,D.I.: Filtering of semi-martingales with counting observations. Proc. Sixth Brasov Conference on Prob. Theory, September 1979 (to appear).

[6] Hadžiev, D.I.: On the predictable expectation with respect to a point measure. Proc. Tenth Spring Conference of the Union of Bulgarian Mathematicians, Sunny Beach, April 1981 (to appear).

Institut of Mathematics
Bulgarian Academy of Sciences
P.O. Box 373, 1090- Sofia,
Bulgaria

EXISTENCE OF PARTIALLY OBSERVABLE STOCHASTIC OPTIMAL CONTROLS*

U.G. Haussmann

Department of Mathematics
University of British Columbia
Vancouver, B.C., V6T 1W5, Canada

1. Introduction.

Consider the following control problem. Minimize

(1.1) $$J[u] = E\{\int_0^T f_o(t,X_t,Y_t,u_t)dt + g_o(X_T,Y_T)\}$$

where $T < \infty$ is fixed, u is in a set $\mathcal{U}$, and

(1.2) $$dX_t = [f(t,X_t,Y_t) + g(t,X_t,Y_t)u_t]dt + \sigma(t,X_t,Y_t)dw_t ,$$

(1.3) $$dY_t = h(t,X_t,Y_t)dt + d\bar{w}_t , \quad Y_o = 0 .$$

For convenience only we assume that X_o is fixed. Here w, $\bar{w}$ are independent $\mathbb{R}^q$ and $\mathbb{R}^m$ valued Wiener processes on a probability space with filtration $(\Omega,\mathcal{F},P,\{\mathcal{F}_t\}_{0\le t\le T})$, $X_t \in \mathbb{R}^d$ is the process to be controlled, $Y_t \in \mathbb{R}^m$ is the observation process, where $\mathbb{R}^m$ is m-dimensional Euclidean space. Traditionally the set $\mathcal{U}$ of admissible controls is taken to be the feedback controls depending on the past of the observation, and assuming values in some set $U \subset \mathbb{R}^p$. Unfortunately the existence of an optimal u in this class, called the strictly admissible controls, is an open question; however recently Fleming and Paradoux [4] showed that an optimal control exists in a wider class of controls, the randomized controls. These controls are actually probability distributions on the space of (Y_t,u_t) trajectories, i.e. the joint probability distributions of the observation and and control processes.

We prove here the same result; however we prefer to define randomized controls slightly differently, as stochastic processes depending on an extra random variable. Also we relax the smoothness hypotheses made in [4], i.e. f, g need not be Lipschitz in X , and h need not be twice continuously differentiable. The proof in [4] uses p.d.e. theory as well as some recent results from filtering theory. We use a simple version of a proof devised by Kushner [7] to solve the completely observable problem. Our proof can also be generalized to cases where (1.2) is not linear in u , c.f. [5]. Since it is lengthy we only sketch it here and leave details to the above reference.

*This work was supported by NSERC under grant A 8051.

Other results where the class $\mathcal{U}$ is much more restricted, but where an optimal strictly admissible control exists can be found in [1], [5], [6]

In the next section we state the assumptions and result precisely, and then in section three sketch the proof.

2. The Result.

We make the following assumptions.

(A_1) U is compact, convex.

(A_2) f,h,g,σ are Borel measurable functions of (t,x,y), $\mathbb{R}^d$ valued, $\mathbb{R}^m$ valued, $d\times p$ matrix valued and $d\times q$ matrix valued respectively, continuous in (x,y) a.e.t.,

$$|f(t,x,y)| + |h(t,x,y)| + |g(t,x,y)| \le K(1+|x|+|y|)$$
$$|\sigma(t,x,y)| \le K ,$$

for some constant K.

(A_3) f_o and g_o are scalar valued, continuous functions of (x,y,u) and (x,y) respectively a.e.t, such that

$$|f_o(t,x,y,u)| + |g_o(x,y)| \le K(1+|x|^r+|y|^r)$$

for some $r < \infty$, and for each (t,x,y), $f_0(t,x,y,\cdot)$ is convex on U.

We write $C^m[0,T]$ for the space of continuous functions mapping $[0,T]$ into $\mathbb{R}^m$, and we let $\mathcal{G}$ be the Borel algebra on $C^m[0,T]$, $\{\mathcal{G}_t\}$ be the canonical filtration of sub σ-algebras. $\mathcal{B}$ is the Borel σ-algebra on $[0,T]$. Then u is a strictly admissible control if $u : [0,T] \times C^m[0,T] \to U$ is $\mathcal{B}\times\mathcal{G}$ measurable, $\{\mathcal{G}_t\}$ adapted, and if there exists a probability space with filtration $(\Omega,\mathcal{F},P,\{\mathcal{F}_t\}_{0\le t\le T})$ carrying independent standard Wiener processes w, $\bar{w}$, and $\{\mathcal{F}_t\}$ adapted processes X, Y such that (1.2) (1.3) hold with $u_t = u(t,Y.)$.

Let us write $L_2([0,T];U)$ or just L_2 for the Lebesgue space of square integrable functions mapping $[0,T]$ into U. Since U is compact and convex, then L_2 under the weak topology is a complete, separable, metrizable space. On L_2 we shall put the Borel σ-algebra induced by the weak topology, and denote it by $\mathcal{B}_L$.

A randomized control u is an adapted stochastic process on some probability space with filtration $(\Omega,\mathcal{F},P,\{\mathcal{F}_t\}_{0\le t\le T})$. Moreover the probability space must carry independent Wiener processes w, $\bar{w}$, adapted processes X, Y satisfying (1.2), (1.3), and a $\mathcal{K}$-valued ($\mathcal{K}$ is a metric space) random variable ζ. This last is the randomization parameter. u should have some further properties: it is completely determined by the observation and ζ, i.e. if $\mathcal{F}^{Y\zeta}$ is the sub σ-algebra of $\mathcal{F}$ generated by Y and ζ, then

(i) $u : \Omega \to L_2$ is $(F^{Y\zeta}, B_L)$ measurable.

Furthermore u_t should be independent of w and of the future observations. Unfortunately this independence holds not on (Ω, F, P), but rather on a new probability space defined as follows. Let

(2.1) $$Z_t = \int_0^t h(s,X_s,Y_s)^* dY_s - \frac{1}{2}\int_0^t |h(s,X_x,X_s)|^2 ds ,$$

where $*$ denotes transpose, and let Q be the measure given by

(2.2) $$\frac{dQ}{dP} = \exp(-Z_T) .$$

Then w and Y are independent Wiener processes on $(\Omega, F, Q, \{F_t\}_{0\le t\le T})$. Now u is assumed to satisfy

(ii) $\{(Y_r,u_r) : r \le t\}$ and $\{Y_s - Y_t : s \ge t\}$ are independent with respect to Q.

(iii) (Y,u) and w are independent with respect to Q.

Finally the randomization should be on u only, so a randomized control must also satisfy in addition to (i), (ii), (iii):

(iv) Y and ζ are independent with respect to Q.

(v) $(X,Y,w,\bar{w})$ and ζ are conditionally independent with respect to P given (Y,u). We denote the set of randomized controls by U.

Some remarks are in order.

(a) The growth conditions imply that

$$\sup_u \sup_t E(|X_t|^r + |Y_t|^r) < \infty$$

if $E|X_o|^r < \infty$. But X_o is fixed so

(2.3) $$J[u] \equiv E\{\int_0^T f_o(t,X_t,Y_t,u_t)dt + g_o(X_T,Y_T)\} < \infty .$$

(b) If ζ is in fact constant (i.e. no randomization) then (i) and (ii) imply that u is adapted to Y, i.e. u is strictly admissible. In general (i) and (ii) do <u>not</u> even imply that u_t is $F_t^{Y\zeta}$ measurable, where $F_t^{Y\zeta}$ is the σ-algebra generated by ζ and $\{Y_s : s \le t\}$.

(c) If μ is m-dimensional Wiener measure, if p is the distribution of ζ under Q, and if by (i) u is written as

$$u(Y,\zeta)(\cdot) \equiv u_Y(\zeta)(\cdot) \in L_2([0,T];U) ,$$

then π, the joint distribution of (Y,u), is given by

$$\pi(A\times B) \equiv \int_A p[u_y^{-1}(B)]\mu(dy) .$$

In the terminology of [4], π is a randomized control [corresponding to our randomized control u].

We can now state the following existence result.

Theorem. Assume $A_1 - A_3$. There exists a randomized control u^o such that

$$J[u^o] = J^* \equiv \inf\{J[u] : u \in \mathcal{U}\} .$$

We note that in [4] it is shown that $J^* = \inf\{J[u] : u \text{ strictly admissible}\}$ so that u^o also solves the strict sense control problem: $\inf\{J[u] : u \text{ strictly admissible}\}$. This result is true even with our weaker hypotheses and different notion of randomized control provided f, g, σ are Lipschitz in x.

3. The Proof.

We give now an outline of the proof of the theorem. Let $\{u^n\}_{n=1}^{\infty}$ be a minimizing sequence of randomized controls on $(\Omega^n, F^n, P^n, \{F_t^n\}_{0\le t\le T})$, and define Z^n, Q^n as in (2.1), (2.2). Then

$$dX_t^n = [f(t,X_t^n,Y_t^n) + g(t,X_t^n,Y_t^n)u_t^n]dt + \sigma(t,X_t^n,Y_t^n)dw_t^n ,$$

$$J[u^n] = E_Q^n\{e^{Z_t^n}[\int_0^T f_o(t,X_t^n,Y_t^n,u_t^n)dt + g_o(X_T^n,Y_T^n)]\} \to J^*$$

where E_Q^n denotes expectation with respect to Q^n. Now let

$$H_t^n = \int_0^t h(s,X_s^n,Y_s^n)^* dY_s^n : \Omega^n \to C^1[0,T] ,$$

$$B_t^n = \int_0^t \sigma(s,X_s^n,Y_s^n)dw_s^n : \Omega^n \to C^d[0,T] ,$$

$$\Phi_t^n = (w_t^n,X_t^n,Y_t^n,u_t^n,B_t^n,H_t^n) .$$

It can now be shown that the sequence $\{\Phi^n\}$ is tight. The tightness of all but the fourth component of Φ^n follows readily, c.f. [7], and the tightness of $\{u^n\}$ follows from the compactness of $L_2([0,T];U)$. Prohorov's theorem now gives the existence of a weakly convergent subsequence again denoted by $\{\Phi^n\}$, and Skorohod's imbedding lemma implies that we can consider $\Phi^n \to \Phi^o$ w.p.1. on some fixed probability space (Ω,F,Q) for some Φ^o, as in [7]. Moreover by continuity of f and the w.p.1. weak convergence of $u^n(\omega)$ to $u^o(\omega)$ and strong L_2 convergence of $g(\cdot,X^n(\omega), Y^n(\omega))$ to $g(\cdot,X^o(\omega),Y^o(\omega))$

it follows that

$$X_t^o = X_o + \int_0^t [f(s,X_s^o,Y_s^o) + g(s,X_s^o,Y_s^o)u_s^o]ds + B_t^o .$$

A result of Ersov [3] gives the existence of a metric space K and random variable ζ on K such that u^o is $F^{Y\zeta}$ measurable.

If we set $F_t = F_t^{w^o X^o Y^o u^o}$, then as in [7] it can be shown that B^o , H^o are F_t-martingales with representation

$$B_t^o = \int_0^t \sigma(t,X_t^o,Y_t^o)dw_t^o ,$$

$$H_t^o = \int_0^t h(t,X_t^o,Y_t^o)^* dY_t^o .$$

Hence (X^o,Y^o,u^o,w^o) satisfy (1.2), and w.p.1

$$Z_T^n \to Z_t^o \equiv H_T^o - \frac{1}{2}\int_0^T |h(t,X_t^o,Y_t^o)|^2 dt .$$

Finally, since $u^n \to u^o$ weakly in $L_2([0,T] \times \Omega)$ also, then according to [2], v.7.43,

$$u^o \in \bigcap_{n=1}^{\infty} \overline{co}\{u^k : k = n,n+1,\dots\}$$

so that there exists a sequence v^n converging strongly in $L_2([0,T] \times \Omega)$ to u^o with $v^n \in co\{u^k : k=n,n+1,\dots\}$. From the convexity and continuity of f_o and the w.p.1 (hence in probability) convergence of (X^n,Y^n) , it follows that

$$\begin{aligned} & f_o(t,X_t^n,Y_t^n,v_t^n) \\ & \le \sum_{k\ge n} \alpha_k^n f_o(t,X_t^n,Y_t^n,u_t^k) \\ & \le \sum_{k\ge n} \alpha_k^n f_o(t,X_t^k,Y_t^k,u_t^k) + \varepsilon \end{aligned}$$

for $n > n(\varepsilon)$ if $\omega \notin A_\varepsilon$ where $Q(A_\varepsilon) < \varepsilon$. Hence

$$J[u^o] \le J^*$$

because of the uniform integrability of $\{e^{Z_T^n}\}$. On the other hand $u^o \in U$ and $J^* = \inf\{J[u] : u \in U\}$ so equality holds.

References

[1] Christopeit, N., Existence of Optimal Stochastic Controls under Partial Observation , Z. Wahrscheinlichkeits-th. verw. Gebiete, 51(1980), pp.201-213.

[2] Dunford, N. and Schwartz, J.T., Linear Operators, Part I, Interscience, New York, N.Y., 1966.

[3] Ersov, M.P., The Choquet Theorem and Stochastic Equations, Analysis Matematica 1(1975), pp. 259-271.

[4] Fleming, W.H. and Pardoux, E., Existence of Optimal Controls for Partially Observed Diffusions , preprint, Brown University.

[5] Haussmann, U., On the Existence of Optimal Controls for Partially Observed Diffusion, in preparation.

[6] Kohlmann, M., Existence of Optimal Controls for a Partially Observed Semimartingale Control Problem , preprint, Bonn University.

[7] Kushner, H.J., Existence Results for Optimal Stochastic Controls , JOTA, 15(1975), pp. 347-359.

ON THE GENERALIZATION OF THE FEFFERMAN-GARSIA INEQUALITY

By
S. Ishak and J. Mogyoródi
University of Budapest

The purpose of this short note is to generalize the famous Fefferman-Garsia inequality. We shall do this by using the Herz-Garsia inequality, the decomposition of the random variables belonging to the Hardy-space $\mathcal{H}_p$, $1 \le p < +\infty$ and by modifying the notion of the space $\mathcal{K}_q$, due to Garsia, where $2 \le q \le +\infty$.

1. Let $(\Omega, \mathcal{A}, P)$ be a probability space and let $X, Y, \ldots \in L_1(\Omega, \mathcal{A}, P) = L_1$ be random variables. Futher, let $(\mathcal{F}_n \uparrow)$ be an increasing sequence of σ-fields of events, $n \ge 0$. We suppose that the σ-field

$$\mathcal{F}_\infty = \sigma\left(\bigcup_{n=0}^{\infty} \mathcal{F}_n\right)$$

is equal to $\mathcal{A}$. We consider the martingales

$$X_n = E(X|\mathcal{F}_n),\ Y_n = E(Y|\mathcal{F}_n), \ldots,\ n \ge 0,$$

where for the sake of commodity we suppose that $X_o = Y_o = \ldots = 0$ a.e.

2. We say that $X \in L_1$ belongs to $\mathcal{H}_p$, where $1 \le p < +\infty$, if

$$S = S(X) = \left(\sum_{i=1}^{\infty} d_i^2\right)^{\frac{1}{2}} \in L_p$$

Here

$$d_i = X_i - X_{i-1},\ i \ge 1,\ d_o = X_o = 0 \quad \text{a.e.}$$

are the differences of the martingale $(X_n, \mathcal{F}_n)$. If $X \in \mathcal{H}_p$ then we set

$$||X||_{\mathcal{H}_p} = ||S||_p.$$

It is proved that $||X||_{\mathcal{H}_p}$ defines a norm on $\mathcal{H}_p$.

The random variable $X \in \mathcal{H}_p$, $1 \le p < +\infty$, can be decomposed in the following form:

$$X = X' + X'',$$

where X' has the property that $E(X' | \mathcal{F}_0) = 0$ a.e. and

$$\sum_{i=1}^{\infty} |E(X'|\mathcal{F}_i) - E(X'|\mathcal{F}_{i-1})| \in L_p.$$

More precisely,

$$\| \sum_{i=1}^{\infty} |E(X'|\mathcal{F}_i) - E(X'|\mathcal{F}_{i-1})| \|_p \le (4 + 4p) \| X^* \| p.$$

We also have

$$|E(X'|\mathcal{F}_i) - E(X'|\mathcal{F}_{i-1})| \le \delta_i = 4(X_i^* - X_{i-1}^*) + 4\, E(X_i^* - X_{i-1}^* | \mathcal{F}_{i-1}),$$

where

$$X_i^* = \max_{0 \le k \le i} |X_k| \quad \text{and} \quad X^* = \sup_{k \ge 0} |X_k|$$

are the corresponding maximal functions of the martingale $(X_n, \mathcal{F}_n)$.

The second term X'' belongs to the space $\mathcal{P}_p$. This space is defined as follows: we say that $Y \in L_p$ belongs to $\mathcal{P}_p$, where $1 \le p < +\infty$, if it is predictable in L_p, i.e. there exists such a sequence $\{\lambda_n\}$, $n \ge 0$, of random variables for which the following conditions hold:

a) $|Y_n| = |E(Y|\mathcal{F}_n)| \le \lambda_{n-1}$ a.e., $n \ge 1$,

b) $\lambda_0 \le \lambda_1 \le \dots$ a.e.,

c) λ_n is $\mathcal{F}_n$-measurable,

d) $\lambda_\infty = \lim_{n \to +\infty} \lambda_n \in L_p$.

If Y belongs to $\mathcal{P}_p$ then we define

$$\| Y \|_{\mathcal{P}_p} = \inf_{\{\lambda_n\}} \| \lambda_\infty \|_p,$$

where the " inf " is taken for all the L_p-predicting sequences of Y. $\| Y \|_{\mathcal{P}_p}$ is a norm on $\mathcal{P}_p$.

One has shown that $X \in \mathcal{H}_p$ if and only if $X^* \in L_p$. More precisely, this is expressed in the following inequality due to Burkholder, Davis and Gundy

$$c_p||X||_{\mathcal{H}_p} \le ||X^*||_p \le C_p||X||_{\mathcal{H}_p},$$

Here $c_p > 0$ and $C_p > 0$ are constants depending only on p.

Concerning the second member X" in the decomposition of $X \in \mathcal{H}_p$ we have

$$||X''||_{\mathcal{P}_p} \le (13 + 4p)||X^*||_p$$

For the results of this section we refer to [2], Theorem 15.1 and [1], pp 139-141.

3. We say that $Y \in \mathcal{K}_p$ if the set

$$\Gamma_Y^{(p)} = \{\gamma : \gamma \in L_p,\ E(|Y - Y_{n-1}||\mathcal{F}_n) \le E(\gamma|\mathcal{F}_n) \text{ a.e., } \forall n \ge 1\}$$

is not empty. In this case we set

$$||Y||_{\mathcal{K}_p} = \inf_{\gamma \in \Gamma_Y^{(p)}} ||\gamma||_p.$$

Here, $1 \le p < +\infty$. The space $\mathcal{K}_\infty$ is the well-known BMO_1 - space.

We can easily see that $||Y||_{\mathcal{K}_p}$ is a norm on $\mathcal{K}_p$.

4. For the values $2 \le p \le +\infty$ Garsia has introduced the space $\mathcal{K}'_p$. Let $Y \in L_2(\Omega,\mathcal{A},P)$. We say that $Y \in \mathcal{K}'_p$ if the set

$$\Gamma'^{(p)}_Y = \{\gamma : \gamma \in L_p,\ E((Y-Y_{n-1})^2|\mathcal{F}_n) \le E(\gamma^2|\mathcal{F}_n) \text{ a.e., } \forall n \ge 1\}$$

is not empty. In this case we define

$$||Y||_{\mathcal{K}'_p} = \inf_{\gamma \in \Gamma'^{(p)}_Y} ||\gamma||_p.$$

It is easily seen that $||Y||_{\mathcal{K}'_p}$ is a quasi-norm.

Note that the space $\mathcal{K}'_\infty$ is the well-known BMO_2 - space.

It is proved that in case $1 < p < +\infty$ the spaces $\mathcal{K}_p$ and L_p contain the same elements and the norms of these are equivalent. Concerning this assertion we refer to Garsia [1], Theorem III. 5.2., who has shown that $X \in \mathcal{K}_p$ implies that $X^* \in L_p$ and

$$||X^*||_p \le c_p||X||_{\mathcal{K}_p},$$

where $c_p > 0$ is a constant depending only on p.

Also, we can show that $\mathcal{K}'_p$ and $\mathcal{K}_p$ contain the same elements for every $2 \le p \le +\infty$ and that their norms are equivalent.

In fact, let $Y \in \mathcal{K}_p$, where $2 \le p < +\infty$ is any power. Then by the above referred result of Garsia the random variable

$$Y^* = \sup_{n \ge 0} |Y_n|$$

belongs to L_p. Consequently, from the trivial inequality

$$E((Y-Y_{n-1})^2 | \mathcal{F}_n) \le E(2Y^{*2} | \mathcal{F}_n) \text{ a.e.,}$$

it follows that $Y \in \mathcal{K}'_p$. Also, again from the above assertion of Garsia we get

$$||Y||_{\mathcal{K}'_p} \le ||\sqrt{2}\, Y^*||_p \le \sqrt{2}\ c_p ||Y||_{\mathcal{K}_p}.$$

Now suppose that conversely $Y \in \mathcal{K}'_p$. Then as it has been shown by Garsia /cf. [1], Theorem II.1.2./ we have $Y \in \mathcal{H}_p$. This is equivalent to the fact that $Y^* \in L_p$. From this we deduce that $Y \in \mathcal{K}_p$ since trivially

$$E(|Y-Y_{n-1})|\mathcal{F}_n) \le E(2Y^* | \mathcal{F}_n) \text{ a.e.}$$

So, by the Burkholder-Gundy-Davis inequality and the just referred result of Garsia

$$||Y||_{\mathcal{K}_p} \le ||2Y^*||_p \le 2C_p ||Y||_{\mathcal{H}_p} \le 2C_p^{(1)} ||Y||_{\mathcal{K}'_p},$$

where C_p and $C_p^{(1)}$ are positive constants depending only on p.

In view of the facts that $\mathcal{K}'_\infty = BMO_2$ and $\mathcal{K}_\infty = BMO_1$ and that the spaces BMO_1 and BMO_2 are equivalent we also have that $\mathcal{K}_p$ and $\mathcal{K}'_p$ are equivalent for all p such that $2 \le p < +\infty$.

From the above considerations we also deduce that for $2 \le p \le +\infty$ the spaces $\mathcal{K}'_p$ and L_p coincide.

Fefferman and later Garsia have proved the following martingale inequality: if $X \in \mathcal{H}_p$ and $Y \in \mathcal{K}'_q$, where $1 \le p \le 2$ and $q = p/(p-1)$, then the relation

$$|E(X_n Y_n)| \le \sqrt{\frac{2}{p}}\ ||X_n||_{\mathcal{H}_p} ||Y||_{\mathcal{K}'_q}$$

holds and the limit

$$\lim_{n\to+\infty} E(X_n Y_n)$$

is finite. Further, we have

$$|\lim_{n\to+\infty} E(X_n Y_n)| \le \sqrt{\frac{2}{p}}\ ||X||_{\mathcal{H}_p} ||Y||_{\mathcal{K}'_q}.$$

These authors have used this inequality to prove that the dual space of $\mathcal{H}_p$ is $\mathcal{K}'_q$, where $1 \le p \le 2$, $q = p/(p-1)$. In view of the importance of this inequality the authors of the present note have found to be interesting to generalize it for all the values of p such that $1 \le p < +\infty$ and for the spaces $\mathcal{H}_p$ and $\mathcal{K}_q$ instead of the space $\mathcal{K}'_q$.

4. We also define the space $\mathcal{K}^+_p$ as follows: we say that $X \in \mathcal{K}^+_p$ if the set

$$\Delta^{(p)}_X = \{\gamma : \gamma \in L_p,\ E(|X-X_n| \,|\, \mathcal{F}_n) \le E(\gamma|\mathcal{F}_n) \text{ a.e., } \forall n \ge 0\}$$

of random variables is not empty. Here $1 \le p \le +\infty$. If $X \in \mathcal{K}^+_p$ then we set

$$||X||_{\mathcal{K}^+_p} = \inf_{\gamma \in \Delta^{(p)}_X} ||\gamma||_p .$$

$||X||_{\mathcal{K}^+_p}$ defines a norm on the space $\mathcal{K}^+_p$.

The space $\mathcal{K}^+_\infty$ is the well-known BMO^+_1 - space.

The following inequality due to Herz and to Garsia is very beautiful and we shall use it in the sequel.

<u>4.1. Theorem.</u> Let $X \in \mathcal{P}_p$ and $Y \in \mathcal{K}^+_q$, where $1 \le p < +\infty$ and $q = p/(p-1)$. Then for any constant $C > 0$ we have that

$$|E(X_C Y)| \le (4 + 4 \log 2)\ ||X||_{\mathcal{P}_p} ||Y||_{\mathcal{K}^+_q}$$

and that the limit

$$\lim_{C\to+\infty} E(X_C Y)$$

exists.

Here, X_C is a specially constructed bounded random variable tending to X a.e. and in the $\mathcal{P}_p$ - norm.

Concerning this assertion we refer to Garsia [1], pp. 130-137.

To explain this result let us remark that together with X and Y the random variables X_n and Y_n also belong to $\mathcal{P}_p$ and $\mathcal{K}_q^+$, respectively. Further, we have

$$||X_n||_{\mathcal{P}_p} \le ||X||_{\mathcal{P}_p} \text{ and } ||Y_n||_{\mathcal{K}_q^+} \le ||Y||_{\mathcal{K}_q^+}$$

and

$$|X_{n_C}| \le (2 + 2 \log 2)\ C$$

and

$$|X_n - X_{n_C}| \le 2\lambda_{n-1}\ \theta_C(\lambda_{n-1}),$$

where $\{\lambda_n\}$ is an L_p - predicting sequence of X and $\theta_C(\lambda_{n-1})$ is such a random variable for which the inequality

$$0 \le \theta_C(\lambda_{n-1}) \le 1, \quad \forall n \ge 1,$$

holds. Further $\theta_C(\lambda_{n-1})$ converges decreasingly to 0 as $C \to +\infty$.

We can easily show that $\mathcal{K}_p \subset \mathcal{K}_p^+$ and that for $X \in \mathcal{K}_p$ we have

$$||X||_{\mathcal{K}_p^+} \le 2\ ||X||_{\mathcal{K}_p}.$$

The assertion of the preceding theorem leads to the interesting

<u>4.2. Theorem.</u> Let $X \in \mathcal{P}_p$ and $Y \in \mathcal{K}_q$, where $1 \le p < +\infty$ and $q = p/(p-1)$. Then $E(X_nY_n)$ is finite for all $n \ge 1$ and we have

$$|E(X_nY_n)| = |\lim_{C\to+\infty} E(X_{n_C}Y_n)| \le (8 + 8 \log 2)||X_n||_{\mathcal{P}_p}\ ||Y_n||_{\mathcal{K}_q}.$$

<u>Proof.</u> The martingale difference $d_n = X_n - X_{n-1}$ can be estimated in the following way

$$|d_n| = |X_n - X_{n-1}| \le 2\lambda_{n-1},$$

where $\lambda_{n-1} \in L_p$. So, $|d_n| \in L_p$. Also, $Y \in \mathcal{K}_q$ means that the difference

$$d'_n = Y_n - Y_{n-1},\ n \ge 1$$

has the following estimate

$$|d'_n| = E(|Y_n - Y_{n-1}|\mathcal{F}_n) \le E(\gamma|\mathcal{F}_n)$$

The right-hand side here belongs to L_q. So, by the Hölder inequality we get

$$E(|d_i d_j'|) \geq ||d_i||_p ||d_j'||_q \leq 2||\lambda_{i-1}||_p ||\gamma||_q < +\infty.$$

Therefore $E(X_n Y_n)$ is finite and equals

$$\sum_{i=1}^{n} E(d_i d_i').$$

By the Lebesgue dominated convergence theorem we get that

$$\lim_{C\to+\infty} |E((X_n - X_{n_C}) Y_n)| = 0,$$

since by the remarks made before the theorem

$$|X_n - X_{n_C}||Y_n| \leq 2\lambda_\infty \theta_C(\lambda_\infty) |Y_n|$$

and the right-hand side tends decreasingly to 0 as $C\to+\infty$. Further,

$$E(2\lambda_\infty |Y_n|) \leq 2||\lambda_\infty||_p ||Y_n||_q < +\infty.$$

Therefore,

$$E(X_n Y_n) = \lim_{C\to+\infty} E(X_{n_C} Y_n)$$

To prove the last assertion of the theorem we note that according to Theorem 4.1. and the remarks after it the inequality

$$|\lim_{C\to+\infty} E(X_{n_C} Y_n)| \leq (4 + 4 \log 2)||X_n||_{\mathcal{P}_p} ||Y_n||_{\mathcal{K}_q^+} \leq$$

$$\leq (8 + 8 \log 2)||X_n||_{\mathcal{P}_p} ||Y_n||_{\mathcal{K}_q}$$

trivially holds.

This proves the theorem.

5. The random variable $X \in \mathcal{K}_p$, $1 \leq p < +\infty$, can be decomposed in the form

$$X = X' + X'',$$

where X' and X'' have the properties mentioned in section 2.

This enables us to may prove the following.

5.1. Theorem. Suppose that $X \in \mathcal{K}_p$, $1 \leq p < +\infty$, and consider the martingale

$$X_n' = E(X'|\mathcal{F}_n), \ n \geq 0,$$

where X′ is the first term in the decomposition above of $X\in\mathcal{H}_p$. If $Y\in\mathcal{H}_q$, where $q = p/(p-1)$, then for arbitrary $n \geq 1$ we have

$$|E(X'_n\ Y_n)| \leq (4 + 4p)\ ||X^*_n||_p||Y||_{\mathcal{H}_q}$$

and

$$\lim_{n\to+\infty} E(X'_n\ Y_n)$$

exists and is finite. We also have

$$|\lim_{n\to+\infty} E(X'_n\ Y_n)| \leq (4 + 4p)||X^*||_p||Y||_{\mathcal{H}_q}.$$

Proof. Since $Y\in\mathcal{H}_q$ we have that $d'_i = Y_i-Y_{i-1}$, $i \geq 1$, satisfies

$$|d'_i| = E(|d'_i||\mathcal{F}_i) \leq E(\gamma|\mathcal{F}_i),\ i \geq 1,$$

where $\gamma\in\Gamma_Y^{(q)}$ is arbitrary. So, d'_i belongs to L_q and denoting

$$d_i = X'_i - X'_{i-1},\ i \geq 1,$$

we get by the remarks and notations of section 2 that

$$E(|d_i\ d'_j|) \leq E(\delta_i E(\gamma|\mathcal{F}_j)) \leq ||\delta_i||_p||\gamma||_q < +\infty$$

Therefore, for arbitrary $n \geq 0$ we have that $E(X'_n\ Y_n)$ is finite and

$$E(X'_n\ Y_n) = \sum_{i=1}^{n} E(d_i\ d'_i).$$

To show that

$$\lim_{n\to+\infty} E(X'_n\ Y_n)$$

exists and is finite it suffices to prove that $\{E(X'_n\ Y_n)\}$, $n \geq 0$, forms a Cauchy-sequence. For arbitrary $0 \leq m \leq n$ we have from what we proved that

$$|E(X'_n\ Y_n)-E(X'_m\ Y_m)| = |\sum_{i=m+1}^{n} E(d_i\ d'_i)| \leq \sum_{i=m+1}^{n} E(\delta_i\ E(\gamma|\mathcal{F}_i)) =$$

$$= E((\sum_{i=m+1}^{n} \delta_i)\gamma) = 4E\ ((X^*_n - X^*_m)\gamma) + 4E((\sum_{i=m+1}^{n} E(X^*_i - X^*_{i-1}|\mathcal{F}_{i-1}))\gamma).$$

By the inequality of Hölder we get from this

$$|E(X'_n\ Y_n) - E(X'_m\ Y_m)| \leq 4\ ||X^*_n-X^*_m||_p||\gamma||_q +$$

$$+ 4||\sum_{i=m+1}^{n} E(X^*_i-X^*_{i-1}|\mathcal{F}_{i-1})||_p||\gamma||_q.$$

The convexity inequality of Burkholder /[2], Theorem 16.1./ implies that

$$\|\sum_{i=m+1}^{n} E(X_i - X_{i-1}|\mathcal{F}_{i-1})\|_p \le p\|X_n^* - X_m^*\|_p .$$

From these

$$|E(X_n' Y_m) - E(X_m' Y_m)| \le (4 + 4p) \ \|X_n^* - X_m^*\|_p \|\gamma\|_q$$

and the right hand side tends to 0 as m, n → + ∞ since $X_n^* \to X^*$ increasingly and $0 \le X_n^* - X_m^* \le 2 X^*$. So we can use for example the Lebesque dominated convergence theorem.

Finally, taking m = 0 in the preceding inequality we get

$$|E(X_n' Y_n)| \le (4 + 4p)\|X_n^*\|_p \|\gamma\|_q$$

Since this inequality holds for arbitrary $\gamma \in \Gamma_Y^{(q)}$ we get that

$$|E\ (X_n' Y_n)| \le (4 + 4p)\|X_n^*\|_p \|Y\|_{\mathcal{K}_q} .$$

The right-hand side is less than

$$(4 + 4p)\|X_n^*\|_p \|Y\|_{\mathcal{K}_q}$$

So, finally,

$$|\lim_{n\to+\infty} E(X_n' Y_n)| \le (4 + 4p)\|X^*\|_p \|Y\|_{\mathcal{K}_q} .$$

This proves the theorem.

6. We are now in the position to may prove our main assertion.

<u>6.1. Theorem.</u> Let $X \in \mathcal{H}_p$ and $Y \in \mathcal{K}_q$, where $1 \le p < +\infty$ and $q = p/(p-1)$. Then

$$\lim_{n\to+\infty} E(X_n Y_n)$$

exists and is finite. Moreover, we have

$$|E(X_n Y_n)| \le C_p \|X_n\|_{\mathcal{H}_p} \|Y\|_{\mathcal{K}_q} \le C_p \|X\|_{\mathcal{H}_p} \|Y\|_{\mathcal{K}_q} ,$$

where $C_p > 0$ is a constant depending only on p.

<u>Proof.</u> Since $X \in \mathcal{H}_p$ we have the decomposition

$$X = X' + X'',$$

where X' is the random variable of Theorem 5.1. Consequently, for arbitrary $n \geq 1$ we have

$$|E(X'_n Y_n)| \leq (4 + 4p)||X^*_n||_p ||Y||_{\mathcal{K}_q}$$

The random variable X'' satisfies the conditions of Theorem 4.2. Consequently, for arbitrary $n \geq 1$ we have

$$|E(X''_n Y_n)| \leq (8 + 8 \log 2)||X''_n||_{\mathcal{P}_p} ||Y_n||_{\mathcal{K}_q} \leq (8 + 8 \log 2)||X''_n||_{\mathcal{P}_p} ||Y||_{\mathcal{K}_q}$$

In section 2. we have remarked that

$$||X''_n||_{\mathcal{P}_p} \leq (13 + 4p)||X^*_n||_p.$$

The last three inequalities together give

$$|E(X_n Y_n)| \leq ((4 + 4p) + (8 + 8 \log 2)(13 + 4p))||X^*_n||_p ||Y||_{\mathcal{K}_q}.$$

By the remark of section 2. we have

$$||X^*_n||_p \leq C'_p ||X_n||_{\mathcal{H}_p}$$

with some constant C'_p depending only on p. Consequently,

$$|E(X_n Y_n)| \leq (C_p ||X_n||_{\mathcal{H}_p} ||Y||_{\mathcal{K}_q},$$

where

$$C_p = C'_p((4 + 4p) + (8 + 8 \log 2)(13 + 4p)).$$

Finally, we show that the sequence $\{E(X_n Y_n)\}$ has the Cauchy property. To this end let us remark that by the preceding two theorems we have

$$E(X_n Y_n) = \sum_{i=1}^{n} E(d_i d'_i).$$

From this for arbitrary $0 \leq m \leq n$ it follows that

$$|E(X_n Y_n) - E(X_m Y_m)| = |\sum_{i=m+1}^{n} E(d_i d'_i)| = |E((X_n - X_m)Y_n)|$$

and using the just proved inequality we get

$$|E(X_n Y_n) - E(X_m Y_m)| \leq C_p ||X_n - X_m||_{\mathcal{H}_p} ||Y||_{\mathcal{K}_q}.$$

Here the right-hand side tends to 0 as $m \to +\infty$. This proves the theorem.

7. For the values of p such that $1 < p < +\infty$ the assertion of the last theorem is obvious since in this case we have $1 < q < +\infty$ and so by the result of Garsia /cf. [1], Theorem III. 5.2./ we deduce that $Y \in \mathcal{K}_q$ implies $Y \in \mathcal{H}_q$. More precisely,

$$||Y||_{\mathcal{H}_q} \le C_q ||Y||_{\mathcal{K}_q},$$

where $C_q > 0$ is a constant depending only on q.

If $X \in \mathcal{H}_q$ and $Y \in \mathcal{K}_{q'}$ where $1 < p < +\infty$, by the Cauchy-Schwarz inequality we have

$$|E(X_n Y_n)| = |E(\sum_{i=1}^{n} d_i d_i')| \le E(S_n S_n'),$$

where

$$S_n^2 = \sum_{i=1}^{n} d_i^2$$

and

$$S_n'^2 = \sum_{i=1}^{n} d_i'^2 .$$

Consequently, using the inequality of Hölder

$$|E(X_n Y_n)| \le (E(S_n^p))^{\frac{1}{p}} (E(S_n'^q))^{1/q} \le ||X_n||_{\mathcal{H}_p} ||Y||_{\mathcal{K}_q}$$

and then the above inequality between the $\mathcal{H}_q$ and $\mathcal{K}_q$ - norms we get the inequality of 6.1. Theorem:

$$|E(X_n Y_n)| \le C_q ||X_n||_{\mathcal{H}_p} ||Y||_{\mathcal{K}_q}.$$

However the method followed in the present paper can be used in a more general situation, too.

Consider the pair (ϕ, Ψ) of conjugate Young-functions /for the definition and the properties of these we refer to [6]/. For example such a pair is

$$(\frac{X^p}{p}, \frac{X^q}{q}),$$

where $X \ge 0$, $1 < p < +\infty$, $q = \frac{p}{p-1}$. We can define the spaces $\mathcal{H}_\phi$ and $\mathcal{K}_\psi$ for these more general functions, too. We say that $X \in \mathcal{H}_\phi$ if $S \in L^\phi(\Omega, \mathcal{A}, P)$. Here $L^\phi(\Omega, \mathcal{A}, P)$ denotes the Orlicz space generated by ϕ.

Further, let $Y \in L_1(\Omega,\mathcal{A},P)$ be an arbitrary random variable and put

$$\Gamma_Y^{(\Psi)} = \{\gamma : \gamma \in L^{\Psi},\ E(|Y-Y_{n-1}||\mathcal{F}_n) \le E(\gamma|\mathcal{F}_n) \text{ a.e., } \forall n \ge 1\}.$$

We say that $Y \in \mathcal{K}_\Psi$ if $\Gamma_Y^{(\Psi)}$ is not empty and in this case we define

$$||Y||_{\mathcal{K}_\Psi} = \inf_{\gamma \in \Gamma_Y^{(\Psi)}} ||\gamma||_\Psi,$$

$||Y||_{\mathcal{K}_\Psi}$ is a norm on $\mathcal{K}_\Psi$.

The Young-function ϕ is called of finite power if the quantity

$$p = \sup_{X>0} \frac{X\phi'(X)}{\phi(X)},$$

the power of ϕ, is finite. Here $\phi'(X)$ denotes the righthand side derivative of ϕ.

Now, generalizing the result of 6.1. Theorem we can prove

<u>7.1. Theorem.</u> Let $X \in \mathcal{H}_\phi$ and $Y \in \mathcal{K}_\Psi$, where (ϕ,Ψ) is a pair of conjugate Young-functions and suppose that ϕ has finite power. Then we have

$$|E(X_n Y_n)| \le C_\phi\ ||X_n||_{\mathcal{H}_\phi}\ ||Y||_{\mathcal{K}_\Psi},$$

where $C_\phi > 0$ is a constant depending only on ϕ. Further,

$$\lim_{n\to+\infty} E(X_n Y_n)$$

exists and

$$|\lim_{n\to+\infty} E(X_n Y_n)| \le C_\phi\ ||X||_{\mathcal{H}_\phi} ||Y||_{\mathcal{K}_\Psi}$$

The proof of this assertion follows in main lines those of this paper and can be found in [3], [4] and [5].

We note that we have not supposed the finiteness of the power of Ψ. When this is infinite then $\mathcal{K}_\Psi$ is "near" to the space $\mathcal{K}_\infty = BMO_1$ /cf. [7]/.

References

[1] Garsia,A.M.: Martingale inequalities. Benjamin. Reading, Massachusets. 1973.

[2] Burkholder,D.L.: Distribution function inequalities for martingales. Annals of Probability. 1/1973/,

[3-4-5] Ishak,S. and Mogyoródi,J.: On the $\mathcal{P}_\phi$ - spaces and the generalisation of the Herz and Fefferman inequalities. I., II. and III. Submitted to Studia Sci. Math. Hungarica.

[6] Neveu,J.: Discrete parameter martingales. North-Holland. Amsterdam, 1975.

[7] Strömberg,J.-O.: Founded mean oscillation with Orlicz norms and duality of Hardy spaces. Bulletin of the Amer. Math. Soc. /1976/ 953-955.

SOME REMARKS ON THE PURELY NONDETERMINISTIC PROPERTY OF SECOND ORDER RANDOM FIELDS*

G. Kallianpur
University of North Carolina
Chapel Hill, NC 27514 USA

Keywords: Purely nondeterministic, innovation subspaces, weakly stationary, random fields

1. Introduction. The aim of this note is to examine, in a qualitative fashion, a notion of pure nondeterminism and of innovation subspaces for second order random fields. It is motivated by some recent work of Bromley and Kallianpur [1] and of Tjøstheim [8]. For continuous random fields the PND property seems to have been first introduced in [8]. The search for appropriate definitions of these concepts for random fields has attracted some attention in recent years. The approach to "past" and "future" adopted in this note is consistent with that adopted in the theory of multiparameter martingales developed by Wong and Zakai or Cairoli and Walsh (see references in [1]). As explained in the last section, the study of a different, but related problem in [1] has led to the present note. The main point is that the notions of "casuality", nondeterminism etc. for the random fields considered in these papers [1,8] are restricted by the technical requirement of commutativity of certain chains of projection operators. This requirement seems to attach too much importance to the description of the random field in terms of Cartesian coordinates. Other definitions, such as the ones considered in the case of Markov random fields might be more natural or useful.

Although the ideas and definitions to be introduced here apply equally well to generalized random fields, we shall consider only random fields for the sake of simplicity. For our purposes, a random field is a stochastic process depending on d parameters, $d \geq 2$. (We assume $d = 2$ for convenience.)

A continuous second order random field (s.o.r.f.) X is a family of r.v.'s $\{X_{st}(\phi)\}$ where $(s,t) \in \mathbb{R}^2$ and $\phi \in \Phi$, Φ being a Hausdorff space satisfying the second countability axiom. Examples are $X_{st} = (X^1_{st}, \ldots, X^k_{st})$, $\Phi = \mathbb{R}^k$ and $X_{st}(\phi) = \langle X_{st}, \phi \rangle$ where $\phi = \phi^1 e_1 + \ldots + \phi^k e_k$, $\{e_1, \ldots, e_k\}$ being an orthonormal basis in $\mathbb{R}^k$ and $\langle , \rangle$ is the usual inner product. We can also have Hilbert-space valued s.o.r.f.'s by taking Φ to be an infinite dimensional, separable Hilbert-space with inner product $\langle , \rangle$. Then $X_{st}(\phi) = \langle X_{st}, \phi \rangle$, $X_{st} \in \Phi$. The second order property means that $E|X_{st}(\phi)|^2 < \infty$ for all s,t and ϕ. In the Hilbert-space valued case we assume

$E\|X_{st}\|^2 < \infty$ where $\|\cdot\|$ is the norm. For convenience, we set $EX_{st}(\phi) \equiv 0$. It will also be assumed that X is continuous, i.e.,

(i) if $\phi_n \to \phi$ then $E|X_{st}(\phi_n) - X_{st}(\phi)|^2 \to 0$ for every (s,t), and

(ii) for each ϕ, $E|X_{st}(\phi) - X_{s't'}(\phi)|^2 \to 0$ as $(s,t) \to (s',t')$.

The relevant Hilbert spaces for our purposes are the following:

$$L(X) = \overline{sp}\{X_{st}(\phi),\ (s,t) \in \mathbb{R}^2,\ \phi \in \Phi\},$$

$$L(X;\ s,t) = \overline{sp}\{X_{uv}(\phi),\ u \le s,\ v \le t,\ \phi \in \Phi\},$$

$$L^1(X;s) = \overline{sp}\{X_{uv}(\phi),\ u \le s,\ v \in \mathbb{R}^1,\ \phi \in \Phi\},$$

$$L^2(X;t) = \overline{sp}\{X_{uv}(\phi),\ u \in \mathbb{R}^1,\ v \le t,\ \phi \in \Phi\},$$

$$L^1(X;\ -\infty) = \bigcap_s L^1(X;s),\ L^2(X;\ -\infty) = \bigcap_t L^2(X;t).$$

[The Hilbert spaces $L(X)$, $L(X;m,n)$, $L^1(X;m)$, $L^2(X;n)$ (m,n integers) associated with a discrete s.o.r.f. are defined in a similar way.]

Let $P^1(s)$ be the orthogonal projection operator with domain $L(X)$ and range $L^1(X;s+)$ where $L^1(X;s+) = \bigcap_{s'>s} L^1(X;s')$. $P^2(t)$ is similarly defined. Write $\Pi_1 = \{P^1(s)\}$, $s \in \mathbb{R}$ and $\Pi_2 = \{P^2(t)\}$, $t \in \mathbb{R}$. When these projections have to be distinguished from others later on, we shall designate them as $P^{X,1}(s)$, $P^{X,2}(t)$, $\Pi^{X,1}$ and $\Pi^{X,2}$.

The s.o.r.f. X is <u>weakly stationary</u> if its covariance function $E[X_{st}(\phi)\overline{X_{s't'}(\psi)}]$ is a function of $s - s'$, $t - t'$, ϕ and ψ. A continuous, weakly stationary s.o.r.f. X has associated with it two commuting continuous unitary groups $U_1(s)$ and $U_2(t)$ determined on $L(X)$ by the shifts $U_1(s)X_{uv}(\phi) = X_{u+s,v}(\phi)$ and $U_2(t)X_{uv}(\phi) = X_{u,v+t}(\phi)$. It can then be seen that

(1) $U_i(h)\ P^i(a) = P^i(a+h)\ U_i(h)$ and $U_i(h)\ P^j(a) = P^j(a)\ U_i(h)$ for $i,j = 1,2$ $(i \neq j)$.

For a discrete s.o.r.f. $X = \{X_{ij}(\phi),\ (i,j) \in \mathbb{Z}^2,\ \phi \in \Phi\}$, the associated Hilbert spaces, projection operators and the unitary operators $U_i (i=1,2)$ are defined in a similar way. Note that $U_i(m) = U_i^m$ for any integer m and that (1) holds for all integers h and a.

2. <u>Discrete s.o.r.f.'s.</u> The purely nondeterministic (PND) property and the decomposition of a PND random field is most easily described for a discrete s.o.r.f. We need to define first, the following subspaces of $L(X)$. Denoting by Δ_{mn} the cell

with vertices (m-1, n-1), (m-1, n), (m, n-1) and (m,n) (where m and n range over $0, \pm 1, \pm 2, \ldots$), define

$$M(X;\Delta_{mn}) = [L^1(X;m) \ominus L^1(X;m-1)] \cap [L^2(X;n) \ominus L^2(X;n-1)].$$

Clearly, $M(X;\Delta_{mn}) \perp M(X;\Delta_{m',n'})$ if $(m,n) \neq (m',n')$. The following assumption is essential for our purpose:

$$\text{(A)} \quad P^1(m)\, P^2(n) = P^2(n)\, P^1(m) \quad \text{for all integers m and n.}$$

Proposition 1. If condition (A) holds then

$$\text{(1)} \quad L(X) = [L^1(X;-\infty) + L^2(X;-\infty)] \oplus \sum_{m,n} \oplus\, M(X;\Delta_{mn}).$$

In (1) the subspace $L^1(X;-\infty) + L^2(X;-\infty)$ is the direct sum (in general, not the orthogonal direct sum) of $L^1(X;-\infty)$ and $L^2(X;-\infty)$.

In view of Proposition 1 it is natural to define a PND random field as follows:

Definition 1. (PND Property) X is said to be purely nondeterministic if

(i) condition (A) holds and

(ii) $L^1(X;-\infty) + L^2(X;-\infty) = 0$

when 0 represents the trivial subspace consisting of the zero element.

The following definition of the PND property has been given by Tjøstheim in [8]. (Tjøstheim's definition is stated for continuous parameter random fields and for second order generalized random fields.)

Definition 1'. Let $\{\Delta_\alpha\}$ be an arbitrary collection of rectangles (with vertices lying in $\mathbb{Z}_2$) such that $\cup_\alpha \Delta_\alpha = \mathbb{R}^2$. Let $M(X;\Delta_\alpha)$ be defined as above. Then

$$L(X) = \overline{sp}\{\cup_\alpha M(X;\Delta_\alpha)\}.$$

It is easily verified that the two definitions of the PND property are equivalent. In fact, Definition (1) is equivalent to the property

$$L(X) = \sum_{m,n} \oplus\, M(X;\Delta_{mn}).$$

Definition 2. X is deterministic if (A) holds and

$$\text{(2)} \quad L(X) = L^1(X;-\infty) + L^2(X;-\infty).$$

It is clear that for a deterministic random field we must have

$$\sum_{m,n} \oplus M(X;\Delta_{mn}) = 0.$$

In view of the decomposition (1) it seems appropriate to call the subspaces $M(X;\Delta_{mn})$ innovation subspaces, $M(X;\Delta_{mn})$ being the subspace bringing in the "new" information contributed by the cell Δ_{mn}. The terminology seems appropriate for, as we shall see below, the subspaces $M(X;\Delta_{mn})$ have many of the same properties as their counterparts in the case of one parameter processes.

Properties of $M(X;\Delta_{oo})$:

(a) If X is weakly stationary all the subspaces $M(X;\Delta_{mn})$ have the same dimension. From the decomposition of L(X) given by (1) it follows that X is not deterministic if and only if $\dim M(X;\Delta_{oo}) > 0$.

Let $\mathcal{Y} = L^1(X;0) \cap L^2(X;0)$ and $R_i = V_i(\mathcal{Y})$ (i=1,2) where V_i is the isometry obtained by restricting U_i^{-1} to $\mathcal{Y}$. Then

(b) $M(X;\Delta_{oo}) = (\mathcal{Y} \ominus R_1) \cap (\mathcal{Y} \ominus R_2)$.

Proof of (b). We use repeatedly the following two elementary facts:

(i) Let L and M be closed linear subspaces of a Hilbert space H, $M \subset L$. Then $L \ominus M = L \cap M^{\perp}$. Here $L \ominus M = \{h \in L: h \perp M\}$ and $M^{\perp}$ denotes $H \ominus M$.

(ii) L and M are closed linear subspaces of H such that the corresponding orthogonal projections P_L and P_M commute. Then $L = (L \cap M) \oplus (L \cap M^{\perp})$.

Using the properties of U_i (i=1,2) mentioned in the Introduction we have

$$R_1 = V_1[L^1(X;0) \cap L^2(X;0)] = U_1^{-1} L^1(X;0) \cap U_1^{-1} L^2(X;0) = L^1(X;-1) \cap L^2(X;0)$$

and

$$\begin{aligned}
\mathcal{Y} \ominus R_1 &= [L^1(X;0) \cap L^2(X;0)] \cap [L^1(X;-1) \cap L^2(X;0)]^{\perp} \\
&= L^1(X;0) \cap \{L^2(X;0) \cap [L^1(X;-1) \cap L^2(X;0)]^{\perp}\} \\
&= L^1(X;0) \cap [L^2(X;0) \cap L^1(X;-1)^{\perp}] \\
&= L^1(X;0) \cap L^1(X;-1)^{\perp} \cap L^2(X;0) \\
&= [L^1(X;0) \ominus L^1(X;-1)] \cap L^2(X;0).
\end{aligned}$$

Similarly, we have

$$\mathcal{Y} \ominus R_2 = L^1(X;0) \cap [L^2(X;0) \ominus L^2(X;-1)].$$

Hence

$$(\mathcal{Y} \ominus R_1) \cap (\mathcal{Y} \ominus R_2) = [L^1(X;0 \ominus L^1(X;-1)] \cap [L^2(X;0 \ominus L^2(X;-1)]$$

$$= M(X;\Delta_{oo})$$

and (b) is proved.

Using Proposition 1 a decomposition can be obtained for $\mathcal{Y}$. For convenience let us denote $L^1(X;-\infty) + L^2(X;-\infty)$ by L_d. Then, from (1) we obtain

$$\mathcal{Y} = P^1(0)\ P^2(0)\ L(X) = P^1(0)\ P^2(0)\ L_d \oplus \sum \oplus P^1(0)\ P^2(0)\ M(X;\Delta_{mn}).$$

$$P^1(0)\ P^2(0)\ P_{L_d} = P^1(0)\ P^2(0)\ [P^1(-\infty) + P^2(-\infty) - P^1(-\infty)\ P^2(-\infty)]$$

$$= P^1(-\infty)\ P^2(0) + P^1(0)\ P^2(-\infty) - P^1(-\infty)\ P^2(-\infty).$$

Also,

$$P^1(0)\ P^2(0)\ P_{M(X;\Delta_{mn})} = P^1(0)\ P^2(0)\ [P^1(m) - P^1(m-1)][P^2(n)-P^2(n-1)]$$

$$= 0 \quad \text{if } m \geq 1 \text{ or } n \geq 1$$

and $$= P_{M(X;\Delta_{mn})} \quad \text{if } m \leq 0 \text{ and } n \leq 0.$$

Hence we have

(c) $$\mathcal{Y} = [L^1(X;-\infty) \cap L^2(X;0) + L^1(X;0) \cap L^2(X;-\infty)] \oplus \sum_{m\leq 0, n\leq 0} \oplus M(X;\Delta_{mn}).$$

(d) A Wold Decomposition for X. Let X be a second order random field satisfying condition (A). Then

(3) $$X_{mn} = \xi_{mn} + \eta_{mn} \qquad m,n \in \mathbb{Z}$$

where

(i) $\xi \perp \eta$, i.e. $(\xi_{mn}, \xi_{m',n'}) = 0$ for $(m,n) \neq (m',n')$;

(ii) ξ is PND;

(iii) η is deterministic.

The decomposition (3) is the Wold decomposition of X.

Proof. Only the outline need be given. Set $\mathcal{M} = \sum_{m,n} \oplus M(X;\Delta_{mn})$. For any $(m,n) \in \mathbb{Z}_2$ define

$$\xi_{mn} = P_{\mathcal{M}}\, X_{mn} \quad \text{and} \quad \eta_{mn} = P_{\mathcal{M}^{\perp}} X_{mn}$$

where $P_{\mathcal{M}}$ is the orthogonal projection with domain L(X) and range $\mathcal{M}$. The Hilbert spaces $L(\xi)$, $L^1(\xi,m)$, $L^2(\xi;n)$, $L(\eta)$, $L^1(\eta;m)$ and $L^2(\eta;n)$ are defined exactly as for X. Conclusion (i) is obvious. By the definition of ξ, we have $L(\xi) = M$. Let $P^{1,\xi}(m)$ denote the orthogonal projector with domain $L(\xi)$ and range $L^1(\xi;m)$. Define $P^{2,\xi}(n)$ similarly. The commutativity of $P^{1,\xi}(m)$ and $P^{2,\xi}(n)$ (for all integers m and n) is a consequence of the following relations: For all $h \in L(X)$,

$$P^{1,\xi}(m)\ P_M h = P_M P^1(m)\ P_M h \quad \text{and}$$

$$P^{2,\xi}(n)\ P_M h = P_M P^2(n)\ P_M h.$$

Next, for $h \in L(X)$ we have

$$P^{1,\xi}(-\infty)\ P_M h = \lim_{m\to-\infty} P^{1,\xi}(m)\ P_M h = P_M \lim_{m\to-\infty} P^1(m)\ P_M h$$

$$= P_M\ P^1(-\infty)\ P_M h = 0 \quad \text{since } L^1(X;-\infty) \subseteq M^\perp .$$

Similarly, $P^{2,\xi}(-\infty) = 0$. Hence $L^1(\xi;-\infty) = L^2(\xi;-\infty) = 0$ and the PND property of ξ is proved.

<u>Proof of (iii)</u>. From the definition $\eta_{ij} = P_{M^\perp} X_{ij}$ we have

(4) $L^1(\eta;m) = P_{M^\perp} L^1(X;m)$ where

(5) $P_{M^\perp} = P_{L^1(X;-\infty) + L^2(X;-\infty)} = P^1(-\infty) + P^2(-\infty) - P^1(-\infty)\ P^2(-\infty).$

The commutativity of the projectors $P^{1,\eta}(m)$ and $P^{2,\eta}(n)$ is easily shown. From (4) and (5),

(6) $L^1(\eta;m) = P^1(-\infty)\ L^1(X;m) + P^2(-\infty)\ L^1(X;m) - P^1(-\infty)\ P^2(-\infty)\ L^1(X;m).$

$$P^1(-\infty)\ L^1(X;m) = L^1(X;-\infty),\ P^2(-\infty)\ L^1(X;m) = P^1(m)\ P^2(-\infty)\ L(X).$$

Making $m \to -\infty$ in (6) we obtain

$$L^1(\eta;-\infty) = L^1(X;-\infty).$$

Similarly, $L^2(\eta;-\infty) = L^2(X;-\infty)$, and we have

$$L^1(\eta;-\infty) + L^2(\eta;-\infty) = L^1(X;-\infty) + L^2(X;-\infty) = M^\perp = L^1(\eta).$$

Thus η is deterministic and the proof is complete.

3. <u>Continuous s.o.r.f.'s</u>. Once again we make the assumption that the projection families $\Pi^{X,1}$ and $\Pi^{X,2}$ commute: For s,t real,

(B) $P^1(s)\, P^2(t) = P^2(t)\, P^1(s)$.

We also assume that

(C) $L^1(X;-\infty) = L^2(X;-\infty) = 0$.

As in the discrete case, (B) and (C) together constitute the definition of the PND property. By assumption, $\Pi^{X,1}$ and $\Pi^{X,2}$ are one-dimensional spectral measures. The decomposition corresponding to (1) given in Section 2 is now obtained by applying the multiplicity theory of the two-dimensional spectral measure $\mathbb{P}$ determined by $P^1(s)$ and $P^2(t)$. The ideas and work of Plesner and Rokhlin are applicable to spectral projection measures on $\mathbb{R}^d$, $d \geq 1$. (The case considered here is $d = 2$ for simplicity [6].) This approach to the representation theory of single parameter stochastic processes is well known having, in fact, been used in [4] and also in the work of Cramér [2].

Let H be a separable infinite dimensional Hilbert space and $\mathbb{P}$ be a spectral measure on Borel sets in $\mathbb{R}^2$ whose values are orthogonal projection operators on H. It will be assumed that the concepts of spectral type, comparison of spectral types, maximal spectral type, cyclic subspace and cyclic operator are familiar to the reader. (Definitions and relevant details can be found in [4] or [6].) The separability of H implies that there exists a maximal spectral type for $\mathbb{P}$ and the following result is the starting point for the multiplicity theory of X:

Proposition 1. (Plesner-Rokhlin)

(1) $H = \sum_{i=1}^{M} \oplus H_{f_i}$,

where H_{f_i} is the cyclic subspace generated by the element f_i. The decomposition has these properties:

(i) $M \leq \aleph_0$ is uniquely determined (not depending on the choice of the elements f_i).

(ii) The elements f_i may be chosen in different ways but the spectral types ρ_i in the decomposition are uniquely determined. Furthermore, we have

$$\rho_1 > \rho_2 > \ldots > \rho_M .$$

The cardinal number M is defined to be the multiplicity of $\mathbb{P}$.

Returning now to the projection measure $\mathbb{P}$ of X defined above and applying Proposition 1 taking $H = L(X)$ we obtain

$$L(X) = \sum_{i=1}^{M} \oplus H_{f_i} .$$

We shall define M to be the multiplicity of the random field X. From the decomposition (1) we have the following representation: For $(s,t) \in \mathbb{R}^2$ and $\phi \in \Phi$,

$$(2) \quad X_{st}(\phi) = \sum_{i=1}^{M} \int_{-\infty}^{s} \int_{-\infty}^{t} F_i(s,t;\ u,v;\ \phi)\ d\zeta_i$$

where the ζ_i's are mutually orthogonal, orthogonal random measures on $\mathbb{R}^2$ such that

(a) $E\ \zeta_i(\Delta_1)\ \overline{\zeta_j(\Delta_2)} = \delta_{ij}\ \rho_i(\Delta_1 \cap \Delta_2)$ (δ_{ij} is Kronecker's delta),

(b) $\displaystyle\sum_{i=1}^{M} \int_{-\infty}^{s} \int_{-\infty}^{t} \left| F_i(s,t;\ u,v,\phi)\right|^2 d\rho_i(u,v) < \infty$

for every $\phi \in \Phi$ and

(c) $L_+(X;s,t) = \displaystyle\sum_{i=1}^{M} \oplus L_+(\zeta_i;\ s,t)$

where

$$L_+(X;\ s,t) = L(X;s+) \cap L(X;\ t+)$$

and

$$L_+(\zeta_i;\ s,t) = \overline{sp}\{\zeta_i(\Delta),\ \Delta \subset (-\infty,s] \times (-\infty,t]\}.$$

A somewhat different approach to (2) is given in Tjøstheim's paper [8].

The representation (2); (a), (b) and (c) can be taken to be the definition of the PND property of a continuous s.o.r.f. X. It is then easy to see that this is equivalent to the property that $\Pi^{X,1}$ and $\Pi^{X,2}$ are commuting, one dimensional spectral measures.

4. Weakly stationary, PND random fields. If the random field X is weakly stationary it can be shown that the spectral types of the f_i in the representation (2) of Section 3 are Lebesgue. The proof of this fact is easier if one assumes the existence of a spectral density and essentially follows the arguments in Rozanov [7]. Otherwise, one can proceed by generalizing to the two-parameter situation the method of Hanner (see Kallianpur and Mandrekar [4]). The following representation for $X_{st}(\phi)$ is then obtained

$$(1) \quad X_{st}(\phi) = \sum_{i=1}^{M} \int_{-\infty}^{s} \int_{-\infty}^{t} G_i(s,t;\ u,v;\ \phi)\ d\xi_i$$

where the ξ_i's are mutually orthogonal, orthogonal random neasures on $\mathbb{R}^2$ with the following properties [8]:

(2) $E\,\xi_i(\Delta_1)\,\overline{\xi_j(\Delta_2)} = \delta_{ij}\,\mu_2(\Delta_1 \cap \Delta_2)$ where μ_2 is Lebesgue measure in $\mathbb{R}^2$;

(3) $\sum_{i=1}^{M} \int_{-\infty}^{s} \int_{-\infty}^{t} \left|G_i(s,t;\, u,v;\, \phi)\right|^2 d_{\mu_2}(u,v) < \infty$, and the property 2-(c) of Section 3.

For $i = 1,2$ let $\{S^i(h)\}$, $h \geq 0$ be the semigroup obtained by restricting $U^i(-h)$ to $\mathcal{Y} = L(X;\, 0,0)$ and let V^i be the Cayley transform of the infinitesimal generator of $\{S^i(h)\}$. It is reasonable to expect that subspaces $\mathcal{Y} \ominus V^i(\mathcal{Y})$ play a role analogous to the spaces $\mathcal{Y} \ominus R_1$ and $\mathcal{Y} \ominus R_2$ in the description of the innovation subspaces given in Section 2. This point will not be pursued further here.

5. Remarks on the commutativity condition. To study the implication of the condition that the chains of projection operators $\Pi^{X,1}$ and $\Pi^{X,2}$ commute, it is convenient to consider the following set up. Suppose that the random field X can be suitably defined as a "generalized process" $X[f]$ on a separable, Hilbert space H (see Rozanov [7]). For $f \in H$ let

$$E(X[f])^2 < \infty, \quad EX[f] = 0 \quad \text{and}$$

$$E(X[f]\, X[g]) = (Bf,\, g)$$

where $f, g \in H$ and B is the covariance operator of X, i.e., B is a bounded, self-adjoint, positive operator on H. We shall further make two simplifying assumptions:

(i) B has a bounded inverse and

(ii) $\Pi^{X,1}$, $\Pi^{X,2}$ are continuous, i.e., complete chains of orthogonal projectors in the sense of Gohberg and Krein [3].

Let us now introduce families of Hilbert spaces connected with the parametrization of X: Let H_{st}, $(s,t) \in \mathbb{R}^2$ be closed linear subspaces of H with the following properties:

(i) $H_{s't'} \subseteq H_{st}$ whenever $s' \leq s$, $t' \leq t$.

(ii) $\bigvee_{(s,t) \in \mathbb{R}^2} H_{st} = H$.

Write $H_s^1 = \bigvee_{-\infty<t<\infty} H_{st}$, $H_t^2 = \bigvee_{-\infty<s<\infty} H_{st}$, $H_{-\infty}^1 = \bigcap_s H_s^1$ and $H_{-\infty}^2 = \bigcap_t H_t^2$.

(iii) $H^1_{-\infty} = H^2_{-\infty} = 0.$

Let Q^1_s be the $\perp$ projector with range H^1_s and Q^2_t, the $\perp$ projector with range H^2_t. The following further assumptions will be made.

(iv) $Q^1_s\, Q^2_t = Q^2_t\, Q^1_s \quad \forall \quad (s,t) \in \mathbb{R}^2.$

(v) The families of projection operators $\Pi^1 = \{Q^1_s\}$, $s \in \mathbb{R}^1$ and $\Pi^2 = \{Q^2_t\}$, $t \in \mathbb{R}^1$ are complete chains.

The Hilbert spaces corresponding to the random field X are defined in an analogous manner:

$$L(X;\ s,t) = \overline{sp}\{X[f],\ f \in H_{st}\},$$

$$L^1(X;s) = \overline{sp}\{X[f],\ f \in H^1_s\},$$

$$L^2(X;t) = \overline{sp}\{X[f],\ f \in H^2_t\}.$$

The spaces $L(X)$, $L^1(X;-\infty)$ and $L^2(X;-\infty)$ are defined similarly. Let $\Pi^X_1 = \{P^1_s\}$, $\Pi^X_2 = \{P^2_t\}$.

The PND property of X can then be defined in the following abstract form:

There exists an isometry J from H onto L(X) such that

(1) $L(X;\ s,t) = J\, H_{st} \qquad \forall \quad s,t.$

One can then show the following. (Compare [7])

<u>Proposition.</u> Condition (1) holds if and only if the covariance operator B has a special factorization of the following form.

(2) $B = F^*F$

where F is invertible and

(3) $F\, H_{st} = H_{st} \qquad \forall \quad s,t.$

For processes depending on a single parameter the following sufficient condition is known [7]. Let

(4) $B = I - T$, where T is self-adjoint and Hilbert-Schmidt. Then B has the factorization (2) and (3) and so X has the PND property. However, for random fields,

(4) is no longer a sufficient condition for X to have the PND property (1). For the 2-parameter case considered here we have, in general, not the desired factorization (2) but a four-fold special factorization relative to the commuting, complete chains Π_1 and Π_2.

$$(5) \quad B = F_1^* F_2^* F_2 F_1$$

where F_1, F_2 are invertible and F_1 has the property $F_1 H_{st} = H_{st}$. However, (unless $F_2 = 0$), we always have $F_2 H_{st} \subsetneq H_{st}$.

A concrete example of the situation just described was encountered in Bromley and Kallianpur [1] in the study of the problem of finding "non-anticipative" representations of Gaussian processes absolutely continuous with respect to the 2-parameter Wiener process. The factorization (5) is derived and its further properties discussed in [1]. A very brief description of the problem studied in [1] and recast in the framework of this section is as follows. The reader will find the details in [1]. Let $W = (W_{st})$, $0 \le s, t \le 1$ be a 2-parameter Wiener process on a probability space (Ω, F, P). Let $L(W)$ be the linear (Hilbert) space of W and $L(W; s,t) = \overline{sp}\{W_{uv}, 0 \le u \le s, 0 \le v \le t\}$. Consider a Gaussian field X obtained by a linear transformation of W whose probability distribution is absolutely continuous with respect to 2-parameter Wiener measure. To conform to the present set up we define X as a "generalized" process by $X[f] = Af$ where $f \in L(W)$ and A is a bounded, linear operator on $L(W)$. Write $H = L(W)$, $H_{st} = L(W; s,t)$ and $X_{st} = X[W_{st}]$. Then the conditions (i), (iii), (iv) are satisfied (with (iii) modified in the obvious way) and $L(X; s,t)$ is the same as $\overline{sp}\{X_{uv}, 0 \le u \le s, 0 \le v \le t\}$. If S denotes the equivalence operator on $L(W)$ (see [1]) and A is chosen such that $A^*A = S$, then the relation

$$EX[f]\, X[g] = (Sf, g)$$

shows that S is the covariance operator of the generalized process $X[\cdot]$ as defined above. The main theorem of [1] translates into the following result:

X has the PND property in the sense of (1) if and only if the special factorization (2) holds for S. Furthermore, this happens if and only if $F_2 = 0$ in the four-fold factorization of S given by (5).

In the light of this example, the condition of commutativity would appear to be a stringent restriction. In fact, [1] contains instances where commutativity fails to hold. In many problems of practical interest, the particular PND property considered here does not seem to be natural. Further study, particularly, the derivation of sufficient conditions for commutativity (of reasonable generality and usefulness) should clarify the situation.

For stationary random fields, one expects a similar discussion as in this section but with the covariance operator replaced by the spectral density operator.

REFERENCES

[1] Bromley, C. and G. Kallianpur (1980), Gaussian random fields, *Appl. Math. Optim.* 6, 361-376.

[2] Cramér, H. (to appear in 1981), A note on multiplicity theory, *Statistics and Probability*: Essays in Honor of C. R. Rao. (North Holland)

[3] Gohberg, I. C. and M. G. Krein (1969), Theory and applications of Volterra operators in Hilbert space, Vol. 24, *Translations of Mathematical Monographs, American Mathematical Society.*

[4] Kallianpur, G. and V. Mandrekar (1965), Multiplicity and representation theory of purely non-deterministic stochastic processes, *Theory of Probability and Its Applications*, 10, 614-644.

[5] Kallianpur, G. and V. Mandrekar (1966), Semi-groups of isometries and the representation and multiplicity of weakly stationary stochastic processes, *Arkiv, för Matematik*, 6, No. 19, 319-335.

[6] Plesner, A. I. (1969), *Spectral Theory of Linear Operators*, Vol. 2 (Translated by M. K. Nestell and A. G. Gibbs), Frederick Unger Publishing Co., New York.

[7] Rozanov, Yu. A. (1977), *Innovation Processes* (English Translation), V. H. Winston and Sons.

[8] Tjøstheim, D. (1978), Multiplicity theory for random fields using Quantum Mechanical methods, *Probability Theory on Vector Spaces*, Lecture Notes in Mathematics, No. 656, Springer-Verlag.

(*Partial support of NSF)

THE HÖLDER CONTINUITY OF HILBERT SPACE VALUED STOCHASTIC INTEGRALS WITH AN APPLICATION TO SPDE

Peter Kotelenez

Forschungsschwerpunkt Dynamische Systeme
Universität Bremen

Abstract

Necessary and sufficient conditions for Hölder continuity of Hilbert space valued stochastic integrals are given in terms of the associated increasing process. As an application we show that the mild solution of a stochastic evolution equation has a continuous version if the semigroup governing this equation is analytic.

AMS 1979 Primary: 60 H 05, 60 H 15; Secondary: 60 G 17

0. Introduction

$(\Omega, \mathfrak{A}, P)$ shall denote a complete probability space, E the mathematical expectation. The state spaces will be real separable Hilbert spaces K, H with inner products $\langle\cdot,\cdot\rangle_K$ and $\langle\cdot,\cdot\rangle_H$ respectively. $L(K,H)$ is the space of bounded linear operators from K to H. w_t is a K-valued Wiener process with infinitesimal covariance operator W (cf [3]). F_t is a filtration of σ-algebras, which is compatible with $w(t)$ in the usual way. $||\cdot||_H$, $||\cdot||_{L(K,H)}$ etc. denote the norms in the corresponding spaces.

Now let f be a mapping from [a,b], a < b, a, b ∈ R, into some metric space (X,ρ) and $\mu \in [0,1]$. We shall say that f has Hölder modulus μ if

$$\beta(f,b,\mu-\varepsilon) := \sup_{a\le s<s'\le b} \frac{\rho(f(s'),f(s))}{(s'-s)^{\mu-\varepsilon}} < \infty$$

and if $(f,b,\mu+\varepsilon) = \infty$ for all $\varepsilon > 0$. If we cannot decide whether f has Hölder modulus μ or in (0.1) we have only $\beta(f,b,\mu-\varepsilon) < \infty$ for all $\varepsilon > 0$, we shall say f has Hölder modulus $\underline{\mu}$.

In the first part we consider $x_t := \int_a^t \Phi_u dw_u$ where $t \in [a,b]$ and Φ_u is an $L(K,H)$-valued F_u-adapted process such that

$\int_a^b E||\Phi_u||^2_{L(K,H)} du < \infty$. The associated increasing process is given by $A_t := \int_a^t tr(\Phi_u W \Phi_u^*) du$ (cf. [1], [4], [6]). In Theorem 1 we show that $x(\omega)$ has Hölder modulus $\frac{\mu}{2}(\omega)$ if and only if $A(\omega)$ has Hölder modulus $\mu(\omega)$, where the notation "(ω)" indicates that the moduli in general depend on ω . In particular, if $A(\omega)$ has Hölder modulus 0 , then, $x(\omega)$ has also modulus 0 , i.e., is not Hölder continuous. Thus, quite a lot of stochastic integrals with Wiener process as integrator are easily seen not to be Hölder continuous which refutes a fairly widespread viewpoint (cf., e.g., [5], p. 2).

As an application we give in the second part a sufficient condition for the existence of a continuous version of the mild solution of a stochastic evolution equation if the semigroup governing this equation is analytic which improves a result obtained in [6].

All symbols introduced are standing notation throughout the paper.

1. The Hölder continuity of H-valued stochastic integrals

Theorem 1

$\beta(A(\omega),b,\underline{\mu}(\omega)) < \infty$ if and only if $\beta(x(\omega),b,\frac{\underline{\mu}}{2}(\omega)) < \infty$. In particular, $x(\omega)$ has Hölder modulus $\frac{\underline{\mu}}{2}(\omega)$ if and only if $A(\omega)$ has Hölder modulus $\underline{\mu}(\omega)$, where $x_t := \int_a^t \Phi_u dw_u$ and $A_t := \int_a^t tr(\Phi_u W \Phi_u^*) du$.

Proof

To simplify notation we may without loss of generality assume $[a,b] = [0,1]$. The main tools of the proof are Itô's formula to reduce the infinite dimensional problem to a one-dimensional one (s.(1.3)), a martingabe estimate which generalizes the corresponding estimates in Levy's original proof for the modulus of continuity of the real valued Wiener process (s.(1.4) and cf. [8]),[9], an induction procedure, and standard cutting techniques to make the Hölder estimates uniform in ω . Since in its essential parts ((1.3), (1.4), (1.6), (1.7)) the proof works obviously equally well in both directions we shall only prove the Hölder continuity of the stochastic integral under the assumption that the increasing process is Hölder continuous.

(i) By introducing a stopping time $\tau(h)$ such that $A_{\tau(h)} = h$ (cf. [8]) the problem reduces to showing

$$\overline{\lim_{|h-s|\downarrow 0}} \frac{||x_{\tau(h)}(\omega) - x_{\tau(s)}(\omega)||_H}{(h-s)^{\frac{1}{2}-\varepsilon}} = 0 \quad \forall\ \varepsilon > 0, \tag{1.1}$$

since from (1.1) we obtain

$$\overline{\lim_{|h-s|\downarrow 0}} \frac{||x_h(\omega) - x_s(\omega)||_H}{[A_h(\omega) - A_s(\omega)]^{\frac{1}{2}-\varepsilon}} = 0 \quad \forall\ \varepsilon > 0,$$

and, consequently,

$$\overline{\lim_{|h-s|\downarrow 0}} \frac{||x_h(\omega) - x_s(\omega)||_H}{(h-s)^{(\frac{\mu}{2}(\omega)-2\varepsilon)\ \vee\ 0}} \le \overline{\lim_{|h-s|\downarrow 0}} \frac{(A_h(\omega)-A_s(\omega))^{(\frac{1}{2}-\varepsilon)\vee 0}}{(h-s)^{(\frac{\mu}{2}(\omega)-2\varepsilon)\ \vee\ 0}} < \infty$$

$\forall\ \varepsilon > 0$, with $a \vee b := \max(a,b)$, (1.2)

where the last inequality follows from our assumption. Hence we can assume $A_t \equiv t$.

(ii) By stopping we can replace x_t by $x_t^N := \int_0^t \Phi_u^N dw_u$

such that

$$\sup_{0\le t\le 1} ||x_t^N|| \le N \quad \text{a.s. with } N\in\mathbb{N}$$

and

$$P\{\sup_{0\le t\le 1} ||x_t^N - x_t||_H > 0\} \longrightarrow 0, \text{ as } N \longrightarrow \infty .$$

(iii) By Itô's formula (cf. [3]) we obtain

$$\left.\begin{aligned} ||x_{t+s}^N - x_t^N||^2 &= 2 \int_t^{t+s} \langle \Phi_u^{N*}(x_u^N - x_t^N), dw_u\rangle_K \\ &\quad + \int_t^{t+s} \operatorname{tr}(\Phi_u^N W \Phi_u^{N*}) du \end{aligned}\right\} \tag{1.3}$$

$$=: m_{t,t+s}^N + A_{t+s}^N - A_t^N$$

(iv) $m_{t,t+s}^N$ is a real valued continuous martingale in s after t. Hence, for arbitrary $\gamma > 0$ by Itô's formula (cf. [7], [6])

$$\exp(\gamma m_{t,t+s}^N - \gamma^2 2\int_t^{t+s} ||W^{\frac{1}{2}}\Phi_u^{N*}(x_u^N - x_t^N)||_K^2 du) =: Y_{t,t+s}^N$$

is a martingale in s (after t) with $E(y_{t,t+s}) = 1$.

Consequently, for an arbitrary $c > 0$,

$$P\{\sup_{0\le s\le h} m^N_{t,t+s} \ge c\}$$

$$\le P\{\sup_{0\le s\le h} y^N_{t,t+s} \ge \exp(\gamma c - \gamma^2 2\int_t^{t+h} ||W^{\frac{1}{2}}\Phi^N_u * (x^N_u - x^N_t)||^2_K du)\}$$

$$\le P\{\sup_{0\le s\le h} y^N_{t,t+s} \ge \exp(\gamma c - \gamma^2 2(A^N_{t+h} - A^N_t)(\sup_{t\le u\le t+h} ||x^N_u - x^N_t||^2_K))\} \quad (1.4)$$

$$\le P\{\sup_{0\le s\le h} y^N_{t,t+s} \ge \exp(\gamma c - \gamma^2 8N^2 h)\}$$

$$\le E(y^N_{t,t+h})\exp(-\gamma c + 8\gamma^2 N^2 h) \quad \text{by the martingale inequality (cf. [7])}$$

$$= \exp(-\gamma c + 8\gamma^2 N^2 h).$$

Setting $\gamma = \frac{c}{10N^2h}$ the preceding estimates yield

$$P\{\sup_{0\le s\le h} | m^N_{t,t+s} | > c\} \le 2\exp(\frac{-c^2}{50N^2h})$$

If we now take some $L > 50N^2$, $h = 2^{-n}$ and abbreviate

$\alpha := \frac{L}{50N^2}$ we obtain

$$P\{\sup_{0\le s\le 2^{-n}} | m^N_{t,t+s} | \ge (L2^{-n}\log 2^n)^{\frac{1}{2}}\} \le 2\cdot 2^{-\alpha n}$$

wherefrom

$$P\{\sup_{0\le k\le 2^n}\ \sup_{0\le s\le 2^{-n}} | m^N_{k2^{-n},k2^{-n}+s} | \ge (L2^{-n}\log 2^n)^{\frac{1}{2}}\} \le 2\cdot 2^{-(\alpha-1)n}. \quad (1.5)$$

Thus by the Borel-Cantelli Lemma there is a.s. an $n_o(\omega)$ such that for all $n \ge n_o(\omega)$

$$\sup_{0\le k\le 2^n}\ \sup_{0\le s\le 2^{-n}} | m^N_{k2^{-n},k2^{-n}+s} | < (L2^{-n}\log 2^n)^{\frac{1}{2}}. \quad (1.6)$$

Since

$$\sup_{0\le t<t+h\le 1} \frac{A^N_{t+h} - A^N_t}{h} \le \sup_{0\le t<t+h\le 1} \frac{\Lambda_{t+h} - \Lambda_t}{h} = 1$$

by our assumption on A_t we have for all $n \ge n_o(\omega)$

$$\sup_{0\le k\le 2^n}\ \sup_{0\le s\le 2^{-n}} \frac{\|x^N_{k2^{-n}+s} - x^N_{k2^{-n}}\|^2_H}{(2^{-n}\log 2^n)^{\frac{1}{2}}} < \infty \tag{1.7}$$

whence as in Levy's classical proof of the continuity modulus for the real valued Wiener process by representing any two numbers dyadically (cf. [8], p. 16)

$$\sup_{0\le t<t+h\le 1} \frac{\|x^N_{t+h} - x^N_t\|^2_H}{(h\log\frac{1}{h})^{\frac{1}{2}}} . < \infty \quad \text{a.s.} \tag{1.8}$$

(v) Now the proof follows from induction which will be only sketched here:

Again by stopping we replace x^N_t by

$$x^{N,N_1}_t := \int_0^t \Phi^{N,N_1}_u dw_u$$

such that

$$\sup_{0\le t<t+h\le 1} \frac{\|x^{N,N_1}_{t+h} - x^{N,N_1}_t\|^2_H}{(h\log\frac{1}{h})^{\frac{1}{2}}} \le N_1 \quad \text{a.s.}$$

and repeating the estimates of (iv) we obtain (cf. (1.4) and the following formula)

$$P\{\sup_{0\le s\le h} m^{N,N_1}_{t,t+s} > c\} \le P\{\sup_{0\le s\le h} y^{N,N_1}_{t,t+s} \ge \exp(\gamma c-\gamma^2 N_1 h^{1+\frac{1}{2}-\varepsilon_1})\} \tag{1.9}$$

for an arbitrary $\varepsilon_1 > 0$. Hence,

$$\sup_{0\le t<t+h\le 1} \frac{\|x^{N,N_1}_{t+h} - x^{N,N_1}_t\|^2_H}{(h^{1+\frac{1}{\varepsilon}-\varepsilon}\log\frac{1}{h})^{\frac{1}{2}}} < \infty \quad \text{a.s.} \tag{1.10}$$

The rest is obvious. ■

2. A continuous version of the mild solution of stochastic evolution equations of analytic type

We consider the following mild evolution equation for $t \in [t_o,t_1]$

$$\begin{aligned} dz_t &= Az_t dt + B(t,z_t)dw_t + f_t dt \\ z_{t_o} &= z_o \end{aligned} \tag{2.1}$$

where A is the generator of an analytic semigroup of operators T_t, on H, $B(s,x) : [t_o,t_1] \times H \longrightarrow L(K,H)$ is measurable in (s,x) and uniformly Lipschitz in the second component. f_t is an H-valued, F_t-adapted stochastic process which is Bochner integrable in t such that $\int_{t_o}^{t_1} E||f_t||_H^2 dt < \infty$, z_o F_{t_o}-measurable and $E||z_o||_H^2 < \infty$.

Under these conditions (2.1) is known to have a unique mild solution (cf. [1], Theorem 3.2) which satisfies by definition

$$z_t = T_{t-t_o} z_o + \int_{t_o}^{t} T_{t-s} B(s,z_s) dw_s + \int_{t_o}^{t} T_{t-s} f_s ds \tag{2.2}$$

$$\text{a.s.} \quad \text{for} \quad t \in [t_o,t_1] .$$

Theorem 1 yields the following sufficient condition for the existence of a continuous version of (2.2):

Theorem 2

If for one version of (2.2) $\int_{t_o}^{t} \operatorname{tr}(B(s,z_s(\omega))WB^*(s,z_s(\omega))ds$ has Hölder modulus $\mu(\omega) > 0$ a.s., then there is a continuous version of the solution of (2.1).

Proof

The last term in (2.2) can be easily shown to be continuous because of the estimate $||T_t||_{L(H,H)} \leq C$, where $C > 0$ is some constant. However the second term on the r.h.s. in (2.2) is a.s. equal to $x_t + A\int_{t_o}^{t} T_{t-s} x_s ds$, with $x_s = \int_{t_o}^{s} B(u,z_u) dw_u$, after integrating by parts, and $A\int_{t_o}^{t} T_{t-s} x_s(\omega) ds$ is continuous if $x_s(\omega)$ is Hölder continuous - as a purely deterministic problem (cf. [4], lemma 2.30, [2] for the "integration by parts formula", and [6] for more details). ∎

References

[1] Arnold, L.; Curtain, R. F.; Kotelenez, P.
Nonlinear stochastic evolution equations in Hilbert space
Forschungsschwerpunkt Dynamische Systeme
Universität Bremen, Report Nr. 17 (1980)

[2] Chojnowska-Michalik, A.
Stochastic differential equations in Hilbert spaces and their applications
Ph. D. thesis, Institute of Mathematics, Polish Academy of Science, 1976

[3] Curtain, R. F.; Falb, P. L.
Itô's lemma in infinite dimensions
J. math. Analysis Appl. 31, No. 2, 1970, 434-448

[4] Curtain, R. F.; Pritchard, A. J.
Infinite dimensional linear system theory. Lecture notes in control and information sciences Vol. 8, 1978, Springer-Verlag

[5] DaPrato, G.; Lannelli, M.; Tubaro, L.
Some results on linear stochastic differential equations in Hilbert spaces
Libera Universita'Degli Studi Di Trento
UTM 52, September 1979

[6] Kotelenez, P.; Curtain, R. F.
Local behaviour of Hilbert space valued stochastic integrals and the continuity of mild solutions of stochastic evolution equations
Forschungsschwerpunkt Dynamische Systeme
Universität Bremen, Report Nr. 21 (1980)

[7] Liptser, R. S.; Shiryayev, A. N.
Statistics of Random Processes I,
Springer-Verlag, New York 1977

[8] McKean, Jr. H. P.
Stochastic integrals
Academic Press, New York/London 1969

[9] Williams, D.
Diffusions, Markov Processes and Martingales,
Vol. 1, John Wiley & Sons, New York/Toronto 1979

Forschungsschwerpunkt "Dynamische Systeme"
Universität Bremen,
Postfach 330 440, 2800 Bremen 33,
West Germany

ON THE FIRST INTEGRALS AND LIOUVILLE EQUATIONS FOR DIFFUSION PROCESSES

By N.V. Krylov and B.L. Rozovskiĭ

Introduction. Let $(\Omega, \mathcal{F}, P)$ be a complete probability space. Consider an n-dimensional Brownian motion $w(t)$ and a d-dimensional diffusion process $X_{x,s}(t)$ generated by $w(t)$:

$$X_{x,s}(t) = x + \int_s^t b\,(r, X_{x,s}(r))\,dr +$$

$$+ \int_s^t \sigma(r, X_{x,s}(r))\,dw(r) \quad ⊠$$

Definition. The map $Z : [s,T]\times R^d\times\Omega\to R^1$ is a first integral (f.i.) for $X_{x,s}(t)$ if it is a $\mathcal{B}\times\mathcal{F}$-measurable function for every $t\in[s,T]$, where $\mathcal{B}$ is Borel σ-algebra on R^d and $Z(t, X_{x,s}(t)) = Z(t + \tau, X_{x,s}(t + \tau))$ P-a.s. for all $t,\tau \in [s,T]$, $t + \tau\le T$ ⊠

The notion of the f.i. (or motion integral) for dynamic system comes from the analytical mechanics where it plays a very important role (e.g. [1]).

The aim of this paper is to show that this notion is also of interest for the theory of diffusion processes. ⊠

We will consider two types of the f.i.'s, namely backward and forward ones. Say, a f.i. $Z(t,x,\omega)$ is a backward one if it is adapted to the family $\mathcal{F}^t$, a completion of $\sigma\{w(r) - w(\tau),\ r,\ \tau \in [t,T]\}$ with respect to P; and a forward one if it is adapted to the family $\mathcal{F}_t$, a completion of $\sigma\{w(r) - w(\tau),\ r,\tau \in [s,t]\}$ with respect to P.

We derive equations (6), (9) for the forward and backward f.i.'s and call them, keeping to the physical tradition, forward and backward Liouville equations respectively.

Both equations are second order Ito partial differential equations. We derive formula (3) for the solution of these equations, which is a generalization of the famous Feynman-Kac formula.

Obviously, there exists an infinite variety of the f.i. of both types. Two of them are the subject of our special concern. These are backward f.i. $U(t,x)$, determined by boundary condition $U(T,x) = x$ and forward f.i. $V(t,x)$, determined by boundary condition $V(s,x) = x$. It is easy to see that $U(t,x) = X_{x,t}(T)$ and $V(t,x) = X^{-1}_{x,s}(t)$, where $X^{-1}_{\cdot,s}(t)$ is an inverse of the map R^d in R^d given by $X_{\cdot,s}(t)$ (Ito map).

Thus backward Liouville equations for U gives us dynamics of the diffusion particle $X_{x,t}(T)$ with respect to x,t. There is a very strong analogy to the backward Kolmogorov equation which by the way, can be derived from a backward Liouville equation simply by taking expectation. $V(t,x)$ in its turn, presents dynamics of $X^{-1}_{x,s}(t)$ with respect to x,t for a fixed s.

Making use of our generalized Feynman-Kac formula we derive an ordinary Ito equation (13) which describes $X^{-1}_{x,s}(t)$. ⊠

Below we shall systematically use the notion of the backward Ito integral.

<u>Definition.</u> Let $\xi(r)$ be an n-dimensional random process on [O,T], $s \le t$, $s,t \in [O,T]$. Assume

$$\int_s^t \xi(r) * dw(r) = \int_{T-t}^{T-s} \xi(T-r)\, dw_T(r),$$

where $W_T(r) = w(T)-W(T-r)$, if the integral in the right hand side of the equality is defined in the classical sense. It is easy to show that the definition does not depend on the point T. ⊠

Generalization of the Feynman-Kac formula

Consider equation

$$- du(tx) = [tr(a(t,x)\mathcal{D}_{xx}\, u(t,x)) +$$

(1)
$$+ (\mathcal{D}_x\, u(t,x))\, b(t,x) + c(t,x)u(t,x)]\, dt +$$
$$+ [(\mathcal{D}_x u(t,x))d(t,x) + e(t,x)u(t,x)] * dw(t),$$ [1/]

$t \in [s,T)$,

(2) $u(T,x) = \varphi(x)$.

1/ Here and below a, b, d, e are $d\times d$-, $d\times 1$-, $d\times n_1$- and $1\times n_1$-matrixes, respectively, $\mathcal{D}_{xx}$ denotes the matrix of the second derivatives in x, $\mathcal{D}_x$ denotes the row vector of the first derivatives in x and upper index T denotes conjugating.

Let us fix constant $p \geq 2$, integer $m \geq 1$ and make the following assumptions:

A/ functions $a \equiv a^T$, b, c, d, $\mathcal{D}_x d$, $\mathcal{D}_x e^T$ and their derivatives up to the order m (for the function a up to the order $\max(2,m)$) are uniformly (with respect to t,x) bounded;

B/ for all $x, \xi \in R^d$, $t \in [s,T]$

$$\xi^T a(t,x)\xi - \frac{1}{2} ||\xi d(t,x)||^2 \geq 0;$$

C/ $\varphi \in W_p^m \cap W_2^m$ 1/

Let us denote $d\times d$-dimensional square root from $2a-dd^T$ by β. We can and shall assert that elements of β are Lipschitz-continuous in x and uniformly bounded.

Let us also denote $d\times(n+d)$-matrix (d,β) by Σ, and $(n+d)$--dimensional Brownian motion where its first n components coincide with $w(t)$, by $\nu(t)$.

<u>Theorem 1.</u> If the assumptions A/, B/, C/ hold, then the problem (1), (2) has the unique $\mathcal{F}^t$-adapted generalized solution 2/ in the space $L_2(\Omega, C([s,T]; W_2^{m-1}) \cap$

$$\cap L_2(\Omega; C_{weak}([s,T]; W_2^m)) \cap$$

$$\cap L_p(\Omega; C_{weak}([s,T]; W_p^m)) \qquad \text{such that}$$

$$E \sup_{t\in[s,T]} ||u(t)||^p_{W_p^m} < \infty$$

For almost all (t,x) this solution is represented like this

$$u(t,x) = E[\varphi(\xi_{x,t}(T)) \exp\{\int_t^T c(r,\xi_{x,t}(r))\,dr + \tag{3}$$
$$+ \int_t^T e(r,\xi_{x,t}(r))dw(r) - \frac{1}{2}\int_t^T || e(r,\xi_{x,t}(r)||^2\,dr\} | \mathcal{F}^t]$$

1/ W_p^m denotes Sobolev space $W_p^m(R^d)$.

2/ In the sense introduced in [2].

a.s.P, where $\xi_{x,t}(r)$ is the solution of the ordinary Ito equation

$$\xi_{x,t}(r) = x+\int_t^r B(s,\xi_{x,t}(s))\, ds +$$

$$+ \int_t^r \Sigma\, (s,\xi_{x,t}(s))\, d\nu(s),$$

$$B = b - de^T.$$ 1/

Proof. To simplify notations and discussion we prove the statement in the case of $m = \infty$. For the general case see [5]. The first part of the assertion of the theorem follows from Theorem 3 of [6]. Making use of the Sobolev imbedding theorem, it is easy to see that this solution is infinitely differentiable in x, continuous and bounded in (t,x) simultaniously with all its derivatives and satisfies the equation (1) for all x,t a.s.P.

Let L_∞ be the space of all bounded measurable functions h: $[t,T] \to R^n$. Denote

$$q(t) = \exp\{\int_t^T h^T(r)dw(r) - \frac{1}{2}\int_t^T ||h(r)||^2\, dr\}$$

and $\tilde{u} = uq$. If u is the solution of the problem (1), (2) then $\tilde{u}$ is evidently the unique solution of the same type of equation and consequently possesses all of the analytical properties that u has. Thus $v(t,x) = E\tilde{u}(t,x)$ is infinitely differentiable in x, continuous and bounded in (t,x) and is the unique solution in C^2 of the problem

$$-\frac{dv(t,x)}{dt} = \operatorname{tr}\,(a(t,x)\,\partial_{xx}\, v(t,x)) +$$

$$+ (\partial_x\, v(t,x))\,(b(t,x) + d(t,x)\, h(t)) + \tag{4}$$

$$+ (c(t,x) + e(t,x)\, h(t))\, v(t,x),\; t < T,$$

$$v(T,x) = \varphi(x). \tag{5}$$

On the other hand, if u^1 is a right-hand side of the equality (3) $\tilde{u}^1 = u^1 q$ and $v^1 = E\tilde{u}^1$ then, making use of the Girsanov theorem, it is easy to see that

---

1/ A similar formula has been established by H.J.Kushner [3] (formally) and E.Pardoux [4] under some additional assumptions, the most restrictive of which was that the matrix $2a-dd^T$ was supposed to be uniformly non-degenerate. Our proof is completely different and more simple.

$$v^1(t,x) = \tilde{E}\varphi(\xi_{x,t}(T))\exp \int_t^T (c + eh)\ (r,\xi_{x,t}(r))\ dr,$$

where $\tilde{E}$ denotes an expectation with respect to the measure $\tilde{P}$ defined by equality

$$d\tilde{P} = \exp\{ \int_t^T \alpha(r)dw(r) - \frac{1}{2}\int_t^T ||\alpha(r)||^2\ dr\}$$

and $\alpha(r) = e(r,\xi_{x,t}(r)) + h^T\ (r)$.

It also follows from the Girsanov theorem that with respect to the measure $\tilde{P}$, $\xi_{x,t}(r)$ is a solution of the equation

$$\xi_{x,t}(r) = \int_t^r (b + dh)\ (s,\xi_{x,t}(s))\ ds +$$

$$+ \int_t^r \Sigma(s,\xi_{x,t}(s))\ d\tilde{w}(s) + x,$$

where $\tilde{w}(s)$ is an $(n + d)$-dimensional Brownian motion. Consequently, $v^1(t,x)$ is a unique solution of the problem (4), (5). Therefore

$$v^1(t,x) = v(t,x) \qquad \forall(t,x).$$

To complete the proof we should only recall the following statement.

Proposition ([7]). If ζ is $\mathcal{F}^t$-measurable random variable, $E\zeta^2 < \infty$ and $E\zeta q(t) = 0$ for any $h \in L_\infty$, then $\zeta = 0$ P.a.s. ◼

Corollary. Suppose that assumptions of Theorem 1 hold, $m \geq 2$ and u is a generalized solution of the problem (1), (2).

(i) If $\rho \geq 0$ (a.e.), then $u \geq 0$ (a.e.)

(ii) If $e \equiv 0$, $c \leq 0$ and $\varphi \leq 1$ (a.e.) then $u \leq 1$ (a.e.)

Backward and forward Liouville equations. Inverse of the Ito map.

A very simple corollary of Theorem 1 is the following statement.

Theorem 2. Let $b(t,x)$, $\sigma(t,x)$ and their derivatives in x up to the order 3 be uniformly bounded. Then for any $\varphi \in C^2(R^d)$ there exists a unique backward f.i. for $X_{x,s}(t)$ which is a solution (in C^2) of the backward Liouville equation

$$
\begin{aligned}
&- du(t,x) = [\tfrac{1}{2}\, tr(\sigma\sigma^T(t,x)\mathcal{D}_{xx}\, u(t,x)) + \\
(6)\quad &+ (\mathcal{D}_x\, u(t,x))\, b(t,x)]\, dt + \\
&+ (\mathcal{D}_x\, u(t,x))\, \sigma(t,x) * dw(t),\ t \in [s,T)
\end{aligned}
$$

with the boundary condition

(7) $u(T,x) = \varphi(x).$

Furtheremore, the following formula is valid for all t,x a.s.P:

(8) $u(t,x) = \varphi(X_{x,t}(T)).$

<u>Proof.</u> If $\varphi \in C_o^\infty$ then the existence of the generalized solution and formula (8) follows immediately from Theorem 1. On the other hand, making use of formula (8) and well known results (e.g. [8]) on the smoothness of $X_{x,t}(T)$ with respect to x,t, we can conclude that in true fact the solution belongs to C^2. ⊠

<u>Theorem 3.</u> Let b(t,x), σ(t,x) and their derivatives in x up to the order 3 for b and up to the order 4 for σ be uniformly bounded. Then for any $\varphi \in C^2(R^d)$ there exists the unique forward f.i. for $X_{x,s}(t)$ which is the solution (in C^2) of the equation

$$
\begin{aligned}
&du(t,x) = [(\sigma^T(t,x)\mathcal{D}_x{}^T)^T\sigma^T(t,x)\, (\mathcal{D}_x\, u(t,x))^T - \\
&- \tfrac{1}{2}\, tr\, (\sigma\sigma^T\, (t,x)\mathcal{D}_{xx}\, u(t,x)\ - \\
(9)\quad & \\
&- (\mathcal{D}_x\, u(t,x))\, b(t,x)]\, dt - (\mathcal{D}_x\, u(t,x))\, \sigma(t,x) dw(t),\ t > s
\end{aligned}
$$

with the initial condition

(10) $u(s,x) = \varphi(x)$

Furthermore, the following formula is valid for all $(t,x) \in [s,T] \times R^d$ P-a.s.:

(11) $u(t,\ X_{x,s}\ (t)) = \varphi(x).$

Proof. Making change of variables $t = T - \tau + s$ we can transform problem (9), (10) into the problem of the type (6), (7). Then, making use of Theorem 2 we can conclude that this transformed equation has twice continuously differentiable solution u represented like this

$$u(t,x) = \varphi(Y^{x,t}(s)) \tag{12}$$

where $Y^{x,t}(r)$ is the solution of the ordinary backward Ito equation

$$\begin{aligned} Y^{x,t}(r) &= x + \int_r^t (((\sigma^T \mathscr{D}_x^T)^T \sigma^T)^T - b)\ (s, Y^{t,x}(s))\ ds - \\ &- \int_r^t \sigma\ (r, Y^{t,x}(s)) * \ dw(s). \end{aligned} \tag{13}$$

Applying the Ito-Ventcel formula ([9]) to the $u(t, X_{x,s}(t))$ we get by direct calculations that $du(t, X_{x,s}(t)) \equiv 0$ which completes the proof. ⊠

In recent years great attention has been given to the inversion problem for the Ito maps (e.g. [10] and the appended references). The following proposition is our modest contribution to this pot.

Corollary. Under the assumptions of theorem 2 $X_{\cdot,s}(T)$ is of C^2-class diffeomorfism from R^d onto R^d and

$$X^{-1}_{\cdot,s}(T) = Y^{\cdot,T}(s).$$

Proof. It follows obviously from the assertion of Theorem 2 that forward f.i. $V(t,x)$ corresponding to the initial condition $\varphi(x) = x$ is an inverse map to the $X_{\cdot,s}(t)$ and that it belongs to C^2. Moreover, it follows from formula (12) that

$$X^{-1}_{\cdot,s}(t) = V(t,\cdot) = Y^{\cdot,t}(s).$$

To prove that $X_{\cdot,s}(t)$ maps R^d onto R^d it is sufficient to show that inverse map $Y^{\cdot,T}(s)$ is also one to one map. But it follows immediately from.

Theorem 3' Under the assumptions of Theorem 3, there exists the unique forward f.i. for $Y^{x,T}(t)$ which is the solution (in C^2) of the problem (6), (7). Furtheremore, the following formula is valid for all $(t,x) \in [s,T] \times R^d$ P-a.s.

$$u(t, Y^{x,T}(t)) = \varphi(x).$$

We can prove this statement simply by making change of variable $\tau = T + s - t$ in Theorem 2.

Remark. We have stated some kind of duality, namely: backward Liouville equation for $X_{x,s}(t)$ is a forward Liouville equation for its inverse and vice versa. ⊠

Equation (13) for an inverse map was derived independently under a little bit more restrictive assumptions by A.Veretennikov (personal communication).

References

[1] V.I.Arnold: Mathematical methods of classical mechanics, Moscow, "Nauka", 1979 (in Russian).

[2] N.V.Krylov, B.L.Rozovskiǐ: On the Cauchy problem for linear stochastic partial differential equations, Izv.Akad.Nauk SSSR, Ser.Mat.Tom 41 (1977), N 6, pp.1329-1347 (in Russian).

[3] H.J.Kushner: Probability methods for approximations in stochastic control and for elliptic equations, Academic Press, New York, 1977.

[4] E.Pardoux: Stochastic partial differential equations and filtering of diffusion processes. Stochastics, 1979, vol.3, pp.127-167.

[5] N.V.Krylov, B.L.Rozovskiǐ: On the characteristics of the degenerate second order parabolic Ito equations, Proceedings of the I.G.Petrovsky seminar (to appear, in Russian).

[6] B.L.Rozovskiǐ: On conditional distributions of the degenerate diffusion processes, Teor. Veroyatn. i Primen. 1980. N 1, pp. 149-154 (in Russian).

[7] A.K.Zvonkin, N.V.Krylov: On the strong solutions of stochastic differential equations proceedings of the school-seminar on the theory of random processes (Druskininkai, 1974), Part II, Vilnius, 1975, pp.9-88 (in Russian).

[8] Yu.N.Blagoveschensky, M.I.Freidlin: Some properties of diffusion processes depending on a parameter, Soviet Math., Dokl., N 3, 1961, pp.508-511 (in Russian).

[9] B.L.Rozovskiǐ: On the Ito-Ventcel formula, Vestnik of Moscow University, N 1, 1973, pp.26-32 (in Russian).

[10] H.Kunita: On the decomposition of stochastic differential equations (preprint).

N.V. Krilov

Lomonosov University
Department of Probabilites Theory

B.L. Rozovskii

Institute of Higher Qualification
129329 Moscow ul. Kolskaya 2.
U.S.S.R.

AN AVERAGING METHOD FOR THE ANALYSIS OF ADAPTIVE SYSTEMS WITH SMALL ADJUSTMENT RATE

H.J. Kushner
Division of Applied Mathematics
Brown University, Providence, R.I. 02912

ABSTRACT

An averaging method for proving weak convergence of a sequence of non-Markovian processes to a diffusion, together with an averaged Liapunov function stochastic stability technique, are applied to an automata model for route selection in telephone routing. The model is chosen because it is a prototype of a large class to which the methods can be applied. A useful method for applying the basic theorems to such processes is illustrated. Suitably interpolated and normalized "learning or adaptive" processes converge weakly to a diffusion, as the "learning or adaption" rate goes to zero. For small learning rate, the qualitative properties (e.g., asymptotic (large-time) variances and parametric dependence) of the processes can be determined from the properties of the limit. The general approach can be used to study adaptive routing methods in computer and other networks, as well as the asymptotic properties of stochastic difference equations.

I. INTRODUCTION

References [5], [1] develop a useful method to study the asymptotic properties as $\varepsilon \to 0$ and $n\varepsilon \leq T < \infty$ for any real T of solutions to stochastic difference equations of the form

$$(1.1)\qquad Y^{\varepsilon}_{n+1} = Y^{\varepsilon}_{n} + \varepsilon h_{\varepsilon}(Y^{\varepsilon}_{n}, \xi^{\varepsilon}_{n}) + \sqrt{\varepsilon}\, g_{\varepsilon}(Y^{\varepsilon}_{n}, \xi^{\varepsilon}) + o(\varepsilon), \qquad Y^{\varepsilon}_{n} \in R^{r},$$

where the distributions of the random sequence $\{\xi^{\varepsilon}_{n}\}$ might depend on the $\{Y^{\varepsilon}_{n}\}$.

The emphasis in [1] concerned the case where the h_{ε} and g_{ε} are smooth and the convergence was in $D^{r}(0,\infty)$. In many applications in communication, control and automata theory, the h_{ε} and g_{ε} might simply be indicator functions and the noise $\{\xi_{n}\}$ depend on $\{Y^{\varepsilon}_{n}\}$, and the asymptotic properties (as $n \to \infty$, then $\varepsilon \to 0$) desired. Here, we apply the basic results of [5] to one such problem. The problem is typical of a wide class, and the results illustrate the power and applicability of the general technique. The problem here was chosen for specificity, but the method is applicable to a large class of adaptive routing and related problems, as well as to the study of the asymptotic properties of solutions of stochastic finite difference equations.

The basic type of result is the following. Define $Y^\varepsilon(\cdot)$, $t\in[0,\infty)$, by $Y^\varepsilon(0) = Y_0^\varepsilon$ and $Y^\varepsilon(t) = Y_i^\varepsilon$ on $[i\varepsilon, i\varepsilon + \varepsilon)$. Under appropriate conditions, Theorem 1 gives weak convergence of $\{Y^\varepsilon(\cdot)\}$ in $D^r(0,\infty)$ to a particular diffusion process, as $\varepsilon \to 0$. Now let $\{n_\varepsilon\}$ denote a sequence of integers tending to ∞ as $\varepsilon \to 0$. For $t \geq 0$, define $\tilde{Y}^\varepsilon(t) \equiv Y^\varepsilon(t + \varepsilon n_\varepsilon)$. The tilde ~ always denotes a shift by n_ε (discrete parameter) or $\varepsilon n_\varepsilon$ (continuous parameter). By using Theorem 1 on the sequence $\{Y^\varepsilon_{n_\varepsilon+n}, n\geq 0\}$ $= \{\tilde{Y}^\varepsilon_n, n>0\}$ we will get a great deal of information on the asymptotic properties (large n, small ε). The next section gives some background material from [5]. The adaptive routing problem is treated in Sections III to VI.

This paper contains an outline of the basic ideas. Full details are in [7]. The results are the product of a joint effort by the author and Professor Hai Huang of Washington University.

II. A WEAK CONVERGENCE THEOREM

$D^r[0,\infty)$ denotes the space of R^r-valued functions on $[0,\infty)$ which are right-continuous and have left-hand limits, and is endowed with the Skorokhod topology [4]. $\hat{\mathcal{C}}_0$ denotes the continuous functions on $R^r \times [0,\infty)$ with compact support. Let $b_i(\cdot,\cdot)$, $a_{ij}(\cdot,\cdot)$, $i,j \leq r$, be continuous functions on $R^r \times [0,\infty)$ and let

$$A = \sum_i b_i(x,t) \frac{\partial}{\partial x_i} + \frac{1}{2} \sum_{i,j} a_{ij}(x,t) \frac{\partial^2}{\partial x_i \partial x_j}$$

be the infinitesimal operator of a diffusion process $x(\cdot)$. Assume that the solution to the martingale problem (on $D^r[0,\infty)$) of Strook and Varadhan [6] corresponding to A has <u>a unique non-explosive solution for each initial condition</u>.

Let $b_N(\cdot)$ denote a smooth function with values in [0,1], equal to 1 on $S_N = \{x:|x|\leq N\}$, equal to zero in $R^r - S_{N+1}$. Define the truncated sequence $\{Y_n^{\varepsilon,N}, n>0\}$ by

$$(2.1)\quad Y^{\varepsilon,N}_{n+1} = Y^{\varepsilon,N}_n + [\varepsilon h_\varepsilon(Y^{\varepsilon,N}_n, \xi^\varepsilon_n) + \sqrt{\varepsilon} g_\varepsilon(Y^{\varepsilon,N}_n, \xi^\varepsilon_n) + o(\varepsilon)]b_N(Y^{\varepsilon,N}_n) ,$$

$$Y^{\varepsilon,N}_o \equiv Y^\varepsilon_o$$

and define $Y^{\varepsilon,N}(\cdot)$ analogously to $Y^\varepsilon(\cdot)$. It is convenient to state the theorem in terms of $\{Y_n^{\varepsilon,N}\}$ because it is easier to prove tightness of $\{Y^{\varepsilon,N}(\cdot)\}$ than of $\{Y^\varepsilon(\cdot)\}$ directly. Let A^N be the infinitesimal operator of a diffusion process, denoted by $X^N(\cdot)$, whose coefficients $a_N(\cdot,\cdot)$, $b_N(\cdot,\cdot)$ are continuous, bounded, have compact support and equal $a(\cdot,\cdot)$, $b(\cdot,\cdot)$ in S_N. Suppose that $\{Y^{\varepsilon,N}(\cdot)\}$ converges weakly to some such $X^N(\cdot)$ as $\varepsilon \to 0$, for each N. Then [5] $\{Y^{\varepsilon}(\cdot)\}$ converges weakly to $x(\cdot)$ as $n \to \infty$. The following theorem is a restatement of Theorem 3 of [5] with $\tau_\varepsilon = \varepsilon$. Theorem 2 of [5] provides a very convenient method of proving tightness, and we will use it in the sequel. Let $E_n^{\varepsilon,N}$ denote expectation conditioned on $\{Y_j^{\varepsilon,N}, j\leq n, \xi_j^{\varepsilon}, j<n\}$.

Theorem 1. Assume the conditions stated above on the solution to the martingale problem on $D^r[0,\infty)$ corresponding to operator A, and on A^N and $X^N(\cdot)$. For each N, and $f(\cdot,\cdot)\in \mathcal{D}$, a dense set (sup norm) in $\mathcal{L}_0$, let there be a sequence $\{f^{\varepsilon,N}(\cdot)\}$ satisfying the following conditions: it is constant on each interval $[n\varepsilon, n\varepsilon + \varepsilon)$, at $n\varepsilon$ it is measurable with respect to the σ-algebra induced by $\{Y_j^{\varepsilon,N}, \leq n, \xi_j^{\varepsilon}, j<n\}$ and

$$(2.2)\qquad \sup_{n,\varepsilon} E|f^{\varepsilon,N}(n\varepsilon)| + \sup_{n,\varepsilon} \frac{1}{\varepsilon} E|E_n^{\varepsilon,N} f^{\varepsilon,N}(n\varepsilon + \varepsilon) - f^{\varepsilon,N}(n\varepsilon)| < \infty ,$$

and as $\varepsilon \to 0$ and for each t as $n\varepsilon \to t$,

$$(2.3)\qquad E|f^{\varepsilon,N}(n\varepsilon) - f(Y_n^{\varepsilon,N}, n\varepsilon)| \to 0 ,$$

$$(2.4)\qquad E\left|\frac{E_n^{\varepsilon,N} f^{\varepsilon,N}(n\varepsilon + \varepsilon) - f^{\varepsilon,N}(n\varepsilon)}{\varepsilon} - \left(\frac{\partial}{\partial t} + A^N\right) f(Y_n^{\varepsilon,N}, n\varepsilon)\right| \to 0 .$$

Then, if $\{Y^{\varepsilon,N}(\cdot), \varepsilon_0 > \varepsilon > 0\}$ is tight in $D^r[0,\infty)$ for each N, where ε_0 does not depend on N and $Y^{\varepsilon}(0)$ converges weakly to $X(0)$, $\{Y^{\varepsilon}(\cdot)\}$ converges weakly to $X(\cdot)$, the unique solution to the martingale problem with initial condition $X(0)$.

III. AN AUTOMATA PROBLEM - INTRODUCTION

Narendra [2], [3] and others have studied the application of learning theory to problems in the routing of telephone calls through a multinode network. Here, we take one of their models and show how to apply Theorem 1 to get a much more complete asymptotic theory for small rate of change of the automata behavior.

The problem formulation.

Calls arrive at a terminal at random, but only at time instants $n = 0,1,2,\ldots$, with P {one call arrives at nth instant} $= \mu, \mu\in(0,1)$. From the terminal, there are two possible routings, route 1 and route 2, the ith route having N_i independent lines - and can thus handle up to N_i calls simultaneously. Let $[n,n + 1]$ denote the nth interval of time. The duration of each call has a geometric distribution: P {call completed in the $(n + 1)$st interval|uncompleted at end of nth interval, route i used} = λ_i, $\lambda_i \in (0,1)$. The members of the double sequence of the interarrival times and call durations are mutually independent. It is possible to work with more general Markovian arrival and call length processes and with many types of routing mechanisms. In a more general network a vector routing parameter would be used, one component per node. In that case, as in Theorem 3, the average dynamics are used for the stability analysis and then the proof of the appropriate generalization of Theorem 4 would be similar to the proof of Theorem 4.

The parameter ε will be used for the 'rate of adjustment' of the mechanism which selects the route. The routing mechanism works as follows. For each fixed ε, $\{y_n^\varepsilon\}$ denotes a sequence of random variables - with values in $[0,1]$. For definiteness, suppose that the calls terminating in the nth interval actually terminate at time $n + \frac{1}{2}$, and arrivals and route assignments are at the instants $0,1,2,\ldots$ precisely. Define the "route occupancy process" $X_n^\varepsilon = (X_n^{\varepsilon,1}, X_n^{\varepsilon,2})$, where $X_n^{\varepsilon,i}$ is the number of lines of route i occupied at time n^+. If a call arrives at instant $(n + 1)$, the routing mechanism chooses route 1 with probability y_n^ε and chooses route 2 with probability $(1 - y_n^\varepsilon)$. If all lines of the first chosen route i are occupied at instant $(n + 1)^-$, then the call is switched to route $j (j \neq i)$. If all lines of route j are also occupied at instant $(n + 1)^-$, then the call is rejected.

The $\{y_n^\varepsilon\}$ are to be sequentially adjusted so that suitable asymptotic behavior occurs. We use the algorithm (3.1) [3]. Let J_{in}^ε denote the indicator of the event {call arrives at n + 1, is assigned first to route i and is accepted by route i}. For practical purposes, we bound y_n^ε away from the points 0 and 1. Let $0 < y_\ell < y_u < 1$. $\Big|_{y_\ell}^{y_u}$ denotes truncation, and $\alpha(y) = 1 - y$, $\beta(y) = - y$.

$$(3.1) \qquad y_{n+1}^\varepsilon = [y_n^\varepsilon + \varepsilon\alpha(y_n^\varepsilon)J_{1n}^\varepsilon + \varepsilon\beta(y_n^\varepsilon)J_{2n}^\varepsilon]\Big|_{y_\ell}^{y_u} .$$

There are $\alpha_\varepsilon(\cdot) = \alpha(\cdot)$ in $[y_\ell, y_u - \varepsilon]$ and $\beta_\varepsilon(\cdot) = \beta(\cdot)$ in $[y_\ell + \varepsilon, y_u]$ and such that

$$(3.2) \qquad y_{n+1}^\varepsilon = y_n^\varepsilon + [\alpha_\varepsilon(y_n^\varepsilon) J_{1n}^\varepsilon + \beta_\varepsilon(y_n^\varepsilon)J_{2n}^\varepsilon] .$$

Some definitions.

If the choice probabilities y_n^ε are held fixed at some value y for all n, then the route choice mechanism can still be used, but there is no learning. For fixed selection probability $y \in (0,1)$, let $\{X_n(y)\} = \{(X_n^1(y), X_n^2(y))\}, 0 \leq n < \infty\}$ denote the corresponding route occupancy process. For the process $\{X_n(y)\}$, the state space $Z = \{(i,j): i \leq N_1, j \leq N_2\}$ is a single ergodic class, and the probability transition matrix, $A'(y)$, is infinitely differentiable. Define $P_n(\alpha|y) = P\{X_n(y) = \alpha\}$ and define the vector $P_n(y) = \{P_n(\alpha|y), \alpha \in Z\}$. Then

$$(3.3) \qquad P_{n+1}(y) = A(y)\, P_n(y) \quad .$$

The pair $\{(X_n^\varepsilon, y_n^\varepsilon), n \geq 0\}$ is a Markov process on $Z \times [y_\ell, y_u]$. Also

$$(3.4) \qquad P_{n+1}^\varepsilon = A(y_n^\varepsilon)\, P_n^\varepsilon, \text{ where } P_n^\varepsilon = \{P_n^\varepsilon(\alpha), \alpha \in Z\},$$

$$P_n^\varepsilon(\alpha) = P\{X_n^\varepsilon = \alpha | y_\ell^\varepsilon, \ell < n, X_0^\varepsilon\}$$

Let $P(y) = \{P(\alpha|y), \alpha \in Z\}$ denote the unique invariant measure for $\{X_n(y)\}$, and define the stationary probability $P^i(N_i|y) = P\{X_n^i(y) = N_i\}$. Finally, define the transition probability $P(\alpha, j, \alpha_1|y) = P\{X_j(y) = \alpha_1 | X_0(y) = \alpha\}$ and write the marginal transition probability as $P^i(\alpha, j, k|y) = P\{X_j^i(y) = k | X_0(y) = \alpha\}$. Define E_n^ε to be the expectation conditioned on $\{X_\ell^\varepsilon, y_\ell^\varepsilon, \ell \leq n\}$.

A differential equation for the mean value.

Define $\nu_i = (1 - \lambda_i)^{N_i}$. Note that

$$(3.5a) \qquad E_n^\varepsilon J_{1n}^\varepsilon = \mu y_n^\varepsilon [1 - \nu_1 I\{X_n^{\varepsilon,1} = N_1\}] \ ,$$

$$(3.5b) \qquad E_n^\varepsilon J_{2n}^\varepsilon = \mu (1 - y_n^\varepsilon)\, [1 - \nu_2 I\{X_n^{\varepsilon,2} = N_2\}] \quad .$$

For small ε, the behavior of $\{y_n^\varepsilon\}$ is related to the solution of (3.6), where $\hat{F}(y)$ is just $E[\alpha(y)\, J_{1n}^\varepsilon + \beta(y)\, J_{2n}^\varepsilon]$, but with $\{X_n^\varepsilon, y_n^\varepsilon\}$ replaced by $\{X_n(y), y\}$ and using the stationary measure.

$$(3.6) \qquad \dot{y} = \mu\alpha(y)\, y[1 - \nu_1 P^1(N_1|y)] - \mu(1-y)\, \beta(y)[1 - \nu_2 P^2(N_2|y)]$$

$$= \mu y(1 - y)\, [\nu_2 P^2(N_2|y) - \nu_1 P^1(N_1|y)] \equiv \hat{F}(y) \quad .$$

As y increases, $P^1(N_1|y)$ increases (and $P^2(N_2|y)$ decreases) monotonically. Thus, there is a unique point $\bar{y} \in (0,1)$ such that $\hat{F}(\bar{y}) = 0$. Also, $\hat{F}(y) > 0$ for $y < \bar{y}$ and $\hat{F}(y) < 0$ for $y > \bar{y}$. We assume that $\bar{y} \in (y_\ell, y_u)$ and we also assume that $\hat{F}_y(\bar{y}) \neq 0$. For some sequence $n_\varepsilon \to \infty$ as $\varepsilon \to 0$, we study the asymptotic properties of $U^\varepsilon_{n_\varepsilon + n} = (y^\varepsilon_{n_\varepsilon + n} - \bar{y})/\sqrt{\varepsilon} = \tilde{U}^\varepsilon_n = (\tilde{y}^\varepsilon_n - \bar{y})/\sqrt{\varepsilon}$. We let $\tilde{U}^\varepsilon(\cdot)$ denote the piecewise constant interpolation (interval ε) of $\{\tilde{U}^\varepsilon_n\}$. The sequence $\{\tilde{U}^\varepsilon(\cdot)$ converges weakly to the diffusion $u(\cdot)$ defined by (6.3).

IV. SOME PRELIMINARY RESULTS

In this section, we state some auxiliary results concerning uniform convergence of $P_n(y)$ and its derivatives to $P(y)$ and its derivatives. See [7] for the proofs.

Theorem 2. For each $y \in [y_\ell, y_u]$, let $A'(y)$ denote a Markov transition matrix whose components are twice continuously differentiable and such that the corresponding Markov chain $\{X_n(y)\}$ is ergodic with invariant measure $P(y)$. Then $P(\cdot)$ is also continuous and there is a $\delta > 0$ such that the eigenvalues of $A(y)$, except for the single eigenvalue unity, are bounded in absolute value by $1 - \delta$ for all $y \in [y_\ell, y_u]$. $P_n(y)$ converges to $P(y)$ uniformly (and at a geometric rate) in $y \in [y_\ell, y_u]$ and in $P_0(y)$.

The derivatives $P_y(y)$, $P_{yy}(y)$ are continuous and if P_0 does not depend on y, then $P_{n,y}(y)$ and $P_{n,yy}(y)$ converge to $P_y(y)$ and $P_{yy}(y)$, resp, as $n \to \infty$, at a geometric rate which is uniform in P_0 and in $y \in [y_\ell, y_u]$.

V. TIGHTNESS OF $\{U^\varepsilon_n$, SMALL ε, LARGE n$\}$

Theorem 3. There is an $\varepsilon_0 > 0$ such that, for each $\varepsilon < \varepsilon_0$, there is an $N_\varepsilon < \infty$ such that the doubly indexed sequence $\{U^\varepsilon_n, \varepsilon \leq \varepsilon_0\ n \geq N_\varepsilon\}$ is tight, where $U^\varepsilon_n = (y^\varepsilon_n - \bar{y})/\sqrt{\varepsilon}$.

Proof. Define $V(y) = (y - \bar{y})^2$. We have

$$(5.1)\quad E^\varepsilon_n(y^\varepsilon_{n+1} - y^\varepsilon_n) = \mu\varepsilon[\alpha_\varepsilon(y^\varepsilon_n)\, y^\varepsilon_n(1 - \nu_1 I\{X^{\varepsilon,1}_n = N_1\}) + \beta_\varepsilon(y^\varepsilon_n)\,(1 - y^\varepsilon_n)\,(1 - \nu_2 I\{X^{\varepsilon,2}_n = N_2\})] .$$

For small ε,

$$E^\varepsilon_n(y^\varepsilon_n - \bar{y})\,[\alpha_\varepsilon(y^\varepsilon_n)\, J^\varepsilon_{1,n} + \beta_\varepsilon(y^\varepsilon_n)\, J^\varepsilon_{2,n}] \leq E^\varepsilon_n(y^\varepsilon_n - \bar{y})\,[\alpha(y^\varepsilon_n)J^\varepsilon_{1,n} + \beta(y^\varepsilon_n)J^\varepsilon_{2,n}] ,$$

since $0 \leq \alpha_\varepsilon(y) \leq \alpha(y)$ and $\alpha_\varepsilon(y) \neq \alpha(y)$ only if $y_n^\varepsilon - \bar{y} \geq$ for 0 (for small ε), and conversely for the β_ε term. Using the above inequality and $|y_{n+1}^\varepsilon - y_n^\varepsilon| = 0(\varepsilon)$,

$$(5.2)\quad E_n^\varepsilon V(y_{n+1}^\varepsilon) - V(y_n^\varepsilon) \leq 2\mu\varepsilon(y_n^\varepsilon - \bar{y})\,[\alpha(y_n^\varepsilon)\, y_n^\varepsilon(1 - \nu_1 I\{X_n^{\varepsilon,1} = N_1\}) + \beta(y_n^\varepsilon)\,(1 - y_n^\varepsilon)\,(1 - \nu_2 I\{X_n^{\varepsilon,2} = N_2\})] + 0(\varepsilon^2)\ .$$

Define $V_1^\varepsilon(n)$ by

$$(5.3)\quad V_1^\varepsilon(n) = 2\mu\varepsilon(y_n^\varepsilon - \bar{y})\alpha(y_n^\varepsilon)y_n^\varepsilon\nu_1 \sum_{j=n}^{\infty} [P^1(N_1|y_n^\varepsilon) - P^1(X_n^\varepsilon, j-n, N_1|y_n^\varepsilon)] + 2\mu\varepsilon(y_n^\varepsilon - \bar{y})\,\beta(y_n^\varepsilon)\,(1 - y_n^\varepsilon)\nu_2 \sum_{j=n}^{\infty} [P^2(N_2|y_n^\varepsilon) - P^2(X_n^\varepsilon, j-n, N_2|y_n^\varepsilon)].$$

Note that $P^i(X_n^\varepsilon, 0, N_i|y_n^\varepsilon) = I\{X_n^{\varepsilon,i} = N_i\}$. By Theorem 2, the sums converge absolutely and $|V_1^\varepsilon(\cdot)| = 0(\varepsilon)$.

Next, we can show that

$$E_n^\varepsilon V_1^\varepsilon(n+1) - V_1^\varepsilon(n) = -\,2\mu\varepsilon(y_n^\varepsilon - \bar{y})\alpha(y_n^\varepsilon)y_n^\varepsilon\nu_1[P^1(N_1|y_n^\varepsilon) - I\{X_n^{\varepsilon,1} = N_1\}]$$

$$-\,2\mu\varepsilon(y_n^\varepsilon - \bar{y})\beta(y_n^\varepsilon)(1 - y_n^\varepsilon)\ \nu_2[P^2(N_2|y_n) - I\{X_n^{\varepsilon,2} = N_2\}]$$

$$(5.4)\quad + \sum_{j=n+1}^{\infty} 2\mu\varepsilon\nu_1\{E_n^\varepsilon(y_{n+1} - \bar{y})\alpha(y_{n+1}^\varepsilon)P^1(N_1|y_{n+1}^\varepsilon) - P^1(X_{n+1}^\varepsilon, j-n-1, N_1|y_{n+1}^\varepsilon)] - (y_n^\varepsilon - \bar{y})\alpha(y_n^\varepsilon)y_n^\varepsilon[P^1(N_1|y_n^\varepsilon) - P^1(X_n^\varepsilon, j-n, N_1|y_n^\varepsilon)]\}$$

$$+ (\text{a similar sum for route } 2) = 0(\varepsilon^2) \text{ uniformly in } n,\ y_n^\varepsilon,\ X_n^\varepsilon\ [7].$$

In the proof of (5.4), the differentiability result of Theorem 2 and the representation $E_n^\varepsilon P^1(X_{n+1}^\varepsilon, j-n-1, N_1|y_n^\varepsilon) = P^1(X_n^\varepsilon, j-n, N_1|y_n^\varepsilon)$ are used.

Define $V^\varepsilon(n) = V(y_n^\varepsilon) + V_1^\varepsilon(n)$. By (5.2) and (5.4)

$$E_n^\varepsilon V^\varepsilon(n+1) - V^\varepsilon(n) \leq 0(\varepsilon^2) + 2\mu\varepsilon(y_n^\varepsilon - \bar{y})[\alpha(y_n^\varepsilon)y_n^\varepsilon(1 - \nu_1 P^1(N_1|y_n^\varepsilon)) + \beta(y_n^\varepsilon)(1 - y_n^\varepsilon)(1 - \nu_2 P^2(N_2|y_n^\varepsilon))]\ .$$

and there is a $\gamma > 0$ such that

(5.5) $\quad E_n^\varepsilon V^\varepsilon(n+1) - V^\varepsilon(n) \leq O(\varepsilon^2) - \varepsilon\gamma V(y_n^\varepsilon)$.

The existence of the N_ε and the asserted tightness follow from (5.5) and the fact that $|V^\varepsilon(n)| = O(\varepsilon)$ uniformly in n.

Q.E.D.

VI. WEAK CONVERGENCE OF $\{\tilde{U}^\varepsilon(\cdot)\}$

Definition. Recall the definition of N_ε given at the end of the proof of Theorem 3. For any sequence $n_\varepsilon > N_\varepsilon$, define $Q_\varepsilon = n_\varepsilon - N_\varepsilon$. Define $\tilde{J}_{in}^\varepsilon = J_{i,n_\varepsilon+n}^\varepsilon$. Then $\{\tilde{U}_n^\varepsilon\}$ satisfies

(6.1) $\quad \tilde{U}_{n+1}^\varepsilon = \tilde{U}_n^\varepsilon + \sqrt{\varepsilon}[\alpha_\varepsilon(\tilde{y}_n^\varepsilon)\tilde{J}_{1n}^\varepsilon + \beta_\varepsilon(\tilde{y}_n^\varepsilon)\tilde{J}_{2n}^\varepsilon]$.

By Theorem 3, $\{\tilde{U}_n^\varepsilon, \varepsilon \leq \varepsilon_0\}$ is tight. Also, $\tilde{X}_0^\varepsilon = X_{n_\varepsilon}^\varepsilon$.
We now want to prove weak convergence of $\{\tilde{U}^\varepsilon(\cdot)\}$. In order to use Theorem 1, the $\{\tilde{U}_n^\varepsilon\}$ need to be truncated as done in (2.1) to $\{Y_n^\varepsilon\}$. The truncation involves a messier notation, so we use the notation for the untruncated sequence, but we will carry the $b_n(\cdot)$ (see (2.1)) through the calculations. In part 4 of Theorem 4 the 'truncation notation' will be re-introduced. Since for each truncation integer N, $|\tilde{y}_n^\varepsilon - \bar{y}| \leq \sqrt{\varepsilon}(N+1)$, for small ε we can use α, β instead of α_ε, β_ε in (6.1).

We now define some auxiliary processes which are used in the averaging method employed in the proof. Let $\bar{P}$ denote the measure defined by the stationary process $\{X_j(\bar{y}), \infty > j > \infty\}$, with corresponding expectation operator $\bar{E}$. For each n, it is necessary to introduce the process $\{X_j(\bar{y}), j \geq n\}$, but with "initial condition $X_n(\bar{y}) = \tilde{X}_n^\varepsilon$. (I.e., after time n, the route choice probability is $\bar{y}$.) The operator $\bar{E}_n^\varepsilon$ denotes the expectation of functions of this process $\{X_j(\bar{y}), j \geq n\}$ conditional on the "initial" condition $X_n(\bar{y}) = X_n^\varepsilon$. Let $J_{ij}(\bar{y})$ denote the indicator function I{call arrives at j + 1, is assigned to and accepted by route i}, when the route choice variable is $\bar{y}$ and the route occupancy process is $\{X_j(\bar{y})\}$. Whether we intend the ergodic process or the process $\{X_i(\bar{y}), i \geq n\}$ starting at time n with $X_n(\bar{y}) = \tilde{X}_n^\varepsilon$ will be made obvious by use of either $\bar{E}$ or $\bar{E}_n^\varepsilon$. Define

(6.2) $\quad \delta u_j(\bar{y}) = [\alpha(\bar{y})J_{1j}(\bar{y}) + \beta(\bar{y})J_{2j}(\bar{y})]$.

Under $\bar{P}$, the right side has zero expectation.

Theorem 4. For any sequence $n_\varepsilon > N_\varepsilon$, $\{\tilde{U}^\varepsilon(\cdot)\}$ is tight in $D[0,\infty)$. All weakly convergent subsequences converge to a Gauss-Markov diffusion satisfying (6.3). If $\varepsilon Q_\varepsilon \to \infty$ as $\varepsilon \to \infty$, then the limiting diffusion $u(\cdot)$ is stationary in that $u(0)$ has the stationary distribution. (In all cases $u(0)$ is independent of $B(\cdot)$.)

(6.3) $\quad du = Gudt + \sigma dB$, $B(\cdot)$ = standard Brownian motion,

(6.4) $$G = \hat{F}_y(\bar{y}) = \frac{\partial}{\partial y} \mu y(1-y)[\nu_2 P^2(N_2|y) - \nu_1 P^1(N_1|y)]\Big|_{y=\bar{y}},$$

(6.5) $$\sigma^2 = \bar{E}(\delta u_0(\bar{y}))^2 + 2\sum_{n=1}^{\infty} \bar{E}\, \delta u_0(\bar{y})\delta u_n(\bar{y}) .$$

Proof. Part 1. By (5.1),

(6.6) $$\tilde{E}_n^\varepsilon(\tilde{U}_{n+1}^\varepsilon - \tilde{U}_n^\varepsilon) = \sqrt{\varepsilon}\mu\tilde{y}_n^\varepsilon(1-\tilde{y}_n^\varepsilon)[\nu_2 I\{\tilde{X}_n^{\varepsilon,2} = N_2\} - \nu_1 I\{\tilde{X}_n^{\varepsilon,1} = N_1\}]b_N(\tilde{U}_n^\varepsilon) .$$

Let $f(\cdot,\cdot) \in \mathcal{D}$, the space of (x,t) functions with compact support whose mixed partial derivatives up to order 2 in t and 3 in x are continuous. To apply Theorem 1 to $\{\tilde{U}^\varepsilon(\cdot)\}$, we use $f^\varepsilon(\cdot)$ the form

$$f^\varepsilon(n\varepsilon) = f(\tilde{U}_n^\varepsilon, n\varepsilon) + f_0^\varepsilon(n\varepsilon) + f_1^\varepsilon(n\varepsilon) + f_2^\varepsilon(n\varepsilon)$$

where the $f_i^\varepsilon(n\varepsilon)$ will soon be defined. For each N, all $o(\cdot)$ or $O(\cdot)$ are uniform in all variables except their argument. We have

$$\tilde{E}_n^\varepsilon f(\tilde{U}_{n+1}^\varepsilon, n\varepsilon+\varepsilon) - f(\tilde{U}_n^\varepsilon, n\varepsilon) = \tilde{E}_n^\varepsilon[f(\tilde{U}_{n+1}^\varepsilon, n\varepsilon) - f(\tilde{U}_n^\varepsilon, n\varepsilon)]$$
$$+ f_t(\tilde{U}_n^\varepsilon, n\varepsilon)\varepsilon + (\varepsilon) ,$$

$$\tilde{E}_n^\varepsilon[f(\tilde{U}_{n+1}^\varepsilon, n\varepsilon) - f(\tilde{U}_n^\varepsilon, n\varepsilon)] = \tilde{E}_n^\varepsilon f_u(\tilde{U}_n^\varepsilon, n\varepsilon)(\tilde{U}_{n+1}^\varepsilon - \tilde{U}_n^\varepsilon)$$
$$+ \frac{1}{2}\tilde{E}_n^\varepsilon f_{uu}(\tilde{U}_n^\varepsilon, n\varepsilon)(\tilde{U}_{n+1}^\varepsilon - \tilde{U}_n^\varepsilon)^2 + o(\varepsilon)$$

(6.7) $$= \sqrt{\varepsilon}\mu f_u(\tilde{U}_n^\varepsilon, n\varepsilon)\tilde{y}_n^\varepsilon(1-\tilde{y}_n^\varepsilon)b_N(\tilde{U}_n^\varepsilon)[\nu_2 I\{\tilde{X}_n^{\varepsilon,2} = N_1\} - \nu_1 I\{\tilde{X}_n^{\varepsilon,1} = N_1\}]$$
$$+ \frac{f_{uu}(\tilde{U}_n^\varepsilon, n\varepsilon)}{2}\tilde{E}_n^\varepsilon(\tilde{U}_{n+1}^\varepsilon - \tilde{U}_n^\varepsilon)^2 + o(\varepsilon) .$$

By the differentiability result of Theorem 3, we can rewrite the term before the o(ε) as

$$\varepsilon b_N^2(\tilde{U}_n^\varepsilon)\ \frac{f_{uu}(\tilde{U}_n^\varepsilon,\ n\varepsilon)}{2}\ \tilde{E}_n^\varepsilon[\alpha(\tilde{y}_n^\varepsilon)\tilde{J}_{1n}^\varepsilon + \beta(\tilde{y}_n^\varepsilon)\tilde{J}_{2n}^\varepsilon]^2$$

$$= \varepsilon b_N^2(\tilde{U}_n^\varepsilon)\ \frac{f_{uu}(\tilde{U}_n^\varepsilon,\ n\varepsilon)}{2}\ \bar{E}_n^\varepsilon[\alpha(\bar{y})J_{1n}(\bar{y}) + \beta(\bar{y})J_{2n}(\bar{y})]^2 + o(\varepsilon).$$

The terms $\bar{E}_n^\varepsilon J_{1n}(\bar{y})$ and $\tilde{E}_n^\varepsilon J_{1n}^\varepsilon(\tilde{y}_n^\varepsilon)$ differ only in that in the first case $\bar{y}$ is used as the choice variable to get the successor state to $\tilde{X}_n^\varepsilon$, and $\tilde{y}_n^\varepsilon$ is used in the second case.

Part 2. We will "average out" the terms in (6.7) one by one. Define $f_1^\varepsilon(n\varepsilon)$ (analogous to the definition of $V_1^\varepsilon(n)$ in the last section).

$$(6.8)\quad f_1^\varepsilon(n\varepsilon) = \sqrt{\varepsilon}\mu b_N(\tilde{U}_n^\varepsilon)\tilde{y}_n^\varepsilon(1 - \tilde{y}_n^\varepsilon)\ f_u(\tilde{U}_n^\varepsilon,\ n\varepsilon)\ \sum_{j=n}^{\infty}[\nu_2(P^2(\tilde{X}_n^\varepsilon,\ j - n,\ N_2|\tilde{y}_n^\varepsilon)$$

$$- P^2(N_2|\tilde{y}_n^\varepsilon)) - \nu_1(P^1(\tilde{X}_n^\varepsilon,\ j - n,\ N_1|\tilde{y}_n^\varepsilon) - P^1(N_1\ \tilde{y}_n^\varepsilon))]\ .$$

Proceeding analogously to the method of Theorem 3 for $V_1^\varepsilon(n)$, we get [7]

$$(6.9)\quad \tilde{E}_n^\varepsilon f_1^\varepsilon(n\varepsilon + \varepsilon) - f_1^\varepsilon(n\varepsilon)\ \text{-(first term on right of 6.7)} + o(\varepsilon)$$

$$+ \varepsilon\mu b_N(\tilde{U}_n^\varepsilon)\ \frac{\partial}{\partial y}\ \{y(1 - y)[\nu_2P^2(N_2|y) - \nu_1P^1(N_1|y)]\}\tilde{U}_n^\varepsilon\Big|_{y=\bar{y}}$$

$$+ \varepsilon b_N(\tilde{U}_n^\varepsilon)[b_N(\tilde{U}_n^\varepsilon)f_u(\tilde{U}_n^\varepsilon,\ n\varepsilon)]_u \sum_{j=n+1}^{\infty} \bar{E}_n^\varepsilon\delta u_n(\bar{y})\ \delta u_j(\bar{y})\ .$$

Part 3. Now, we "average out" the last term in (6.9). Define $f_2^\varepsilon(n\varepsilon)$ by

$$f_2^\varepsilon(n\varepsilon) = \varepsilon b_N(\tilde{U}_n^\varepsilon)[b_N(\tilde{U}_n^\varepsilon)f_u(\tilde{U}_n^\varepsilon,n\varepsilon)]_u \sum_{j=n}^{\infty}\ \sum_{k=j+1}^{\infty}[\bar{E}_n^\varepsilon\delta u_j(\bar{y})\delta u_k(\bar{y})$$

$$- \bar{E}\delta u_j(\bar{y})\delta u_k(\bar{y})]\ .$$

By the (uniform) geometric convergence result of Theorem 2, the sum converges absolutely and $|f_2^\varepsilon(n\varepsilon)| = O(\varepsilon)$. Using the stationary of

$\{\delta u_n(\bar{y})\}$ under $\bar{P}$, we can show that

$$\tilde{E}_n^\varepsilon f_2^\varepsilon(n\varepsilon + \varepsilon) - f_2^\varepsilon(n\varepsilon) = - \text{(last term of (6.9))} + o(\varepsilon)$$

$$+ \varepsilon b_N(\tilde{U}_n^\varepsilon)[b_N(\tilde{U}_n^\varepsilon) f_u(\tilde{U}_n^\varepsilon, n\varepsilon)]_u \sum_{j=1}^{\infty} \bar{E}(\delta u_0(\bar{y})\delta u_j(\bar{y})) .$$

Finally, we treat the term before the $o(\varepsilon)$ of (6.7) - in the form in which it is written below (6.7). Define $f_0^\varepsilon(n\varepsilon)$ by

$$f_0^\varepsilon(n\varepsilon) = \varepsilon \frac{f_{uu}(\tilde{U}_n^\varepsilon, n\varepsilon)}{2} b_N^2(\tilde{U}_n^\varepsilon) \sum_{j=n}^{\infty} [\bar{E}_n^\varepsilon(\delta u_j(\bar{y}))^2 - \bar{E}(\delta u_j(\bar{y}))^2] .$$

By a procedure similar to that used for $f_1^\varepsilon(n\varepsilon)$, it can readily be shown that

$$\tilde{E}_n^\varepsilon f_0^\varepsilon(n\varepsilon + \varepsilon) - f_0^\varepsilon(n\varepsilon) = o(\varepsilon) + \varepsilon \frac{f_{uu}(\tilde{U}_n^\varepsilon, n\varepsilon)}{2} b_N^2(\tilde{U}_n^\varepsilon)\bar{E}(\delta u_0(\bar{y}))^2$$

$$- \varepsilon \frac{f_{uu}(\tilde{U}_n^\varepsilon, n\varepsilon)}{2} b_N^2(\tilde{U}_n^\varepsilon) E_n^\varepsilon[\alpha(\bar{y})J_{1n}(\bar{y}) + \beta(\bar{y})J_{2n}(\bar{y})]^2.$$

Summarizing the previous calculations

$$\tilde{E}_n^\varepsilon f^\varepsilon(n\varepsilon + \varepsilon) - f^\varepsilon(n\varepsilon) = o(\varepsilon) + \varepsilon f_t(\tilde{U}_n^\varepsilon, n\varepsilon) + \varepsilon f_u(\tilde{U}_n^\varepsilon, n\varepsilon) G U_n^\varepsilon b_N(\tilde{U}_n^\varepsilon)$$

$$\text{(6.10)} \quad + \varepsilon f_u(\tilde{U}_n^\varepsilon, n\varepsilon) b_{N,u}(\tilde{U}_n^\varepsilon) b_N(\tilde{U}_n^\varepsilon) \sum_{j=1}^{\infty} \bar{E}\, \delta u_0(\bar{y})\delta u_j(\bar{y})$$

$$+ \varepsilon \frac{f_{uu}(\tilde{U}_n^\varepsilon, \varepsilon n)}{2} b_N^2(\tilde{U}_n^\varepsilon) [\bar{E}(\delta u_0(\bar{y}))^2 + 2 \sum_{j=1}^{\infty} \bar{E}\, \delta u_0(\bar{y})\delta u_j(\bar{y})] .$$

Part 4. Conclusion. Reintroduce the superscript N. Fix N. All the $f_i^{\varepsilon,N}$ are bounded and of order $o(\sqrt{\varepsilon})$ and $\{\tilde{U}_0^{\varepsilon,N}\} = \{\tilde{U}^{\varepsilon,N}(0)\}$ is tight. Also $\tilde{E}_n^{\varepsilon,N} f^{\varepsilon,N}(n\varepsilon + \varepsilon) - f^{\varepsilon,N}(n\varepsilon) = O(\varepsilon)$. Thus, by [5], Theorem 2, the bounded sequence $\{\tilde{U}^{\varepsilon,N}(\cdot)\}$ is tight in $D[0,\infty)$. Let ε index a weakly convergent subsequence with limit $U^N(\cdot)$. Since A is defined to be the infinitesimal operator of the process $u(\cdot)$ given by (6.3), by (6.10) and Theorem 1, we see that $U^N(\cdot) = u(\cdot)$ until the first escape time from S_N. Thus, by Theorem 1', $\{\tilde{U}^\varepsilon(\cdot)\}$ converges weakly to a solution $u(\cdot)$ of (6.3). The independence of $B(\cdot)$ and $u(0)$ and the stationary assertion are not hard to prove, but we omit the details.

Q.E.D.

ACKNOWLEDGEMENTS

This research has been supported in part by the Air Force Office of Scientific Research (AF-76-3063), the National Science Foundation (Eng. 77-12946), and the Office of Naval Research (N00014-76-C-0279-P003.

REFERENCES

1. Kushner, H.J.., Hai Huang (1979), "On the weak convergence of a sequence of general stochastic difference equations to a diffusion," to appear in SIAM J. on Applied Math.

2. Narendra, K.S., Wright, E.A., Mason, L.E. (1977), "Application of learning automata to telephone traffic routing and control," IEEE Trans. on Systems, Man and Cybernetics, SMC-7, 785-792.

3. Narendra, K.S., Thathachar, M.A.L., (1979), "On the behavior of a learning automaton in a changing environment with application to telephone traffic routing," preprint, Yale University, Dept., of Engineering.

4. Billingsley, P. (1968), Convergence of Probability Measures, John Wiley and Sons, New York.

5. Kushner, H.J. (1979), "A martingale method for the convergence of a sequence of processes to a jump-diffusion process," Z. Wahrscheinlichkeitsteorie, 53, 207-219,(1980).

6. Strook, D.W., Varadhan, S.R.S. (1979), Multidimensional Diffusion Processes, Springer, Berlin.

7. Kushner, H.J., Hai Huang, "Averaging methods for the asymptotic analysis of learning and adaptive systems with small adjustment rate," LCDS Rept. 80-1, April, 1980, Brown University.

Λ-SPACES ASSOCIATED WITH PROCESSES.
APPLICATION TO STOCHASTIC EQUATIONS.

by M. METIVIER
Ecole Polytechnique
Centre de Mathématiques Appliquées
91128 Palaiseau - France -

1 - Λ-SPACES ASSOCIATED TO A PROCESS

1.1 - Λ-spaces

The concept of Λ-space is suggested by different approaches to stochastic integration, in order to propose a unique model in situations apparently as different as the isometric Hilbert valued integral (see 8) and integration with respect to random measures as mentionned below in examples.

Let $\mathbb{B}$ be a Banach space, $\mathbb{H}$ a Hilbert space, $\mathbb{L}$ a closed subspace of the Banach space of bounded linear operators from $\mathbb{B}$ into $\mathbb{H}$ (with the uniform norm) A and $\widetilde{A}$ two positive increasing adapted processes. We consider a vector space Π of processes, the values of which are (possibly unbounded) operators from $\mathbb{B}$ into $\mathbb{H}$, and a mapping λ from Π into the set of positive adapted processes. We denote by $\lambda(\Phi)$ or $(\lambda_t(\Phi))_{t \in \mathbb{R}^+}$ the image of $\Phi \in \Pi$ by λ .

This vector space will be called a <u>Λ-space associated with $\mathbb{L}$, A, $\widetilde{A}$, and the functional λ</u> , if there exists an increasing sequence (τ_n) of stopping times such that $\lim_n \tau_n = +\infty$ a.s. and λ and (τ_n) satisfy together the following properties.

(i) for each $\Phi \in \Lambda$, $(\lambda_t(\Phi))_{t \in \mathbb{R}}$ is a positive adapted process and for each n, $E(\widetilde{A}_{\tau_n^-} \cdot \int_{]0,\tau_n[} \lambda_t(\Phi)\, dA_t) < \infty$

(ii) The mappings $\Phi \sim E(\widetilde{A}_{\tau_n^-} \cdot \int_{]0,\tau_n[} \lambda_t(\Phi)\, dA_t)$, $n \geq 0$ are seminorms on Π giving to Π a structure of complete vector space.

(iii) The set of simple predictable $\mathbb{L}$-valued processes is a dense subspace of Π.

We denote by $\Lambda(\mathbb{L},A,\widetilde{A},\lambda)$ such a Λ-space associated with $\mathbb{L},A,\widetilde{A}$ and λ.

A process X is said to be locally in Λ, if there exists an increasing sequence (σ_n) of stopping times, such that $\lim_n \sigma_n = +\infty$ and the process $1_{]0,\sigma_n[}X$ is in Λ for every n.

1.2 - Λ-stochastic integrals

Let Z be a $\mathbb{B}$-valued regular process. We say that the Λ-space $\Lambda(\mathbb{L},A,\widetilde{A},\lambda)$ is associated with Z if, for every simple predictable $\mathbb{L}$-valued process Y and every stopping time σ the following inequality holds :

$$(1.1)\qquad E(\|\int YdZ^{\sigma}\|^2_{\mathbb{H}}) \leq E(\widetilde{A}_{\sigma-}.\int_{]0,\sigma[}\lambda_s(Y)dA_s)$$

where, by definition, for $Y := \sum_i a_i.1_{]s_i,t_i]\times F_i}$ the $\mathbb{H}$-valued random variable $\int YdZ^{\sigma}$ is given by

$$\int YdZ^{\sigma} := \sum_i 1_{F_i}.a_i(Z^{\sigma}_{t_i} - Z^{\sigma}_{s_i}) \quad \text{and}$$

where Z^{σ} is the process stopped strictly before σ. (See [11], Ch. I)

$$Z^{\sigma}_t(\omega) := \begin{cases} Z_t(\omega) & \text{if } t < \sigma(\omega) \\ Z_{\sigma-}(\omega) & \text{if } t \geq \sigma(\omega) \end{cases}$$

The definition of the Λ-stochastic integral for every process Y which belongs locally to Λ is immediate using continuity.

1.3 - Controlled-process

We say that Z is controlled in $\Lambda(\mathbb{L},A,\widetilde{A},\lambda)$ if for every stopping time τ and every simple $\mathbb{L}$-valued predictable process Y (therefore, for every $Y \in \Lambda(\mathbb{L},A,\widetilde{A},\lambda)$) the following inequality holds :

$$(1.2)\qquad E(\sup_{s<\tau}\|\int_{]0,s]} YdZ\|^2_{\mathbb{H}}) \leq E(\widetilde{A}_{\tau-}.\int_{[0,\tau[}\lambda_s(Y)dA_s).$$

1.4 - Example 1 : Semimartingales and bounded-operator-valued processes

Let Z be an $\mathbb{H}$-valued cad-lag semimartingales ($\mathbb{H}$: Hilbert). We know from [10] that there exists an increasing process (actually many !) such that for every stopping time σ and every $\mathcal{L}_o\mathbb{H}$; $\mathbb{G})_-$ valued predictable process Y we have

$$(1.3)\qquad E\Big(\sup_{0\leq t<\sigma}\Big\|\int_{]0,t]} Y dZ\Big\|^2_{\mathbb{G}}\Big) \leq E\Big(A_{\sigma-}\int_{]0,\sigma[}\|Y_s\|^2\, dA_s\Big)$$

Therefore Z is controlled in $\Lambda(\mathcal{L}(\mathbb{H};\mathbb{G}),A,A,\lambda)$, with $\lambda_s(Y) := \|Y_s\|^2$

1.5 - Example 2 : Isometric integral of unbounded-operator-valued processes

Let M be an $\mathbb{H}$-valued right continuous square integrable martingale.

Let $\tilde{Q}_M$ be the predictable process, whose values are nuclear operators in $\mathbb{H}$ and α_M be the positive measure on predictable sets, as they are defined in [12]. The Hilbert space of processes X, such that $X(\omega,t)$ is for all (ω,t) an operator from $\mathbb{H}$ into $\mathbb{G}$, with domain $D_X(\omega,t) \supset Q_M^{\frac{1}{2}}(\omega,t)$ and

$$\int_{\Omega\times\mathbb{R}^+} \operatorname{trace}(X\circ Q_M \circ Q^\star)\, d\alpha_M < \infty$$

is nothing but the space $\Lambda(\mathcal{L}(\mathbb{H};\mathbb{G}), 1, \langle M\rangle, \lambda)$

$$(1.4)\qquad \text{where } \lambda_s(X) := \operatorname{trace}(X_s \circ Q_M(s) \circ X_s^\star)$$

We have the isometry formula :

$$(1.5)\qquad E\Big\|\int_{]0,\infty[} X dM\Big\|^2 = E\Big(\int_{]0,\infty[}\lambda_1(X)\, d\langle M\rangle_s\Big).$$

If M has no jump on predictable time the process M is also controlled in $\Lambda(\mathcal{L}(\mathbb{H};\mathbb{G}), 1, \langle M\rangle, \lambda)$.

1.6 - Example 3 : Random measures

For every strictly positive bounded function p on the open subset $\underline{E}$ of $\mathbb{R}^d$ let $M^p(\underline{E})$ denote the Banach space such that $\|m\|_p := \int p(x)\,|m|(dx) < \infty$, with the norm $\|m\|_p$.

This is the dual-space of $C^p(\underline{E})$: the space of continuous functions φ on $\underline{E}$ such that

$$\|\varphi\| := \sup_{x \in \underline{E}} \frac{|\varphi(x)|}{p(x)} < \infty \quad .$$

A random measure $\mu(\omega,ds,du)$ of order p is, for each ω a measure $\mu(\omega,ds,.)$ on $\mathbb{R}^+$ with values in M^p. We call F^μ the process (primitive process of μ) defined by $F^\mu_t(\omega) := \mu(\omega,]o,t],.) \in M^p(\underline{E})$. If (F^μ_t) is a martingale we say μ <u>is white</u>.

For every white measure μ there exists a predictable random measure $q(s,\omega,dx \otimes dy)$ on $\underline{E} \times \underline{E}$ of order $p \otimes p$ and an increasing process b with the following properties (see [9]) : for every $\mathbb{H}$-valued "simple" process $(Y(.,s,x))_{s\in\mathbb{R}^+,x\in\underline{E}}$ let us define

$$(1.6) \qquad \lambda_s(Y) := \int_{\underline{E}\times\underline{E}} < Y(.,s,x),Y(.,s,y) >_{\mathbb{H}} q(.,s,dx \otimes dy) .$$

Then $\Lambda(\mathbb{L},1,b,\lambda)$ is a Λ-space associated with F^μ, $\mathbb{L}$ beeing the subspace of $\mathcal{L}(M^p(\underline{E}),\mathbb{H})$ associated with the $\mathbb{H}$-valued "simple" processes $(Y(\omega,s,x))_{s\in\mathbb{R}^+,\, x\in\underline{E}}$.

And moreover for the corresponding Λ-stochastic integral :

$$(1.7) \qquad <\int Y dq>_t = \int_{]0,t]} \lambda_s(Y)\, db(s) .$$

In case of a Poisson point-process with Levy-measure α one has

$$(1.8) \qquad \lambda_s(Y) = \int \|Y(.,s,y)\|^2 \alpha\,(dy) .$$

If F^μ has no jump on predictable stopping times, it is also controlled in $\Lambda(\mathbb{L},1,b,\lambda)$. If F^μ has jump on predictable stopping-times, there exists still a functional of the form (1.6) and an increasing process b such that F^μ is controlled in $\Lambda(\mathbb{L},1,b,\lambda)$, but (1.7) is no larger true for this functional and this process b .

2 - APPLICATION TO STOCHASTIC INTEGRAL EQUATIONS

2.1 - The general equation under consideration

We consider a Banach valued process Z (with values in $\mathbb{B}$). We assume Z controlled in the Λ-space $\Lambda(\mathbb{L},\tilde{A},A,\lambda)$ where $\mathbb{L}$ is a closed subspace of $\mathcal{L}(\mathbb{B};\mathbb{H})$ ($\mathbb{H}$ separable Hilbert space), and V is a regular $\mathbb{H}$-valued process.

The equation under study is the following :

$$(2.1)\qquad \xi_t = V_t + \int_{]0,t]} \underset{\sim}{a}_s \xi \, dZ_s$$

where the functional $\underset{\sim}{a}$ has the following properties :

(i) For every regular $\mathbb{H}$-valued process ξ, $\underset{\sim}{a}\xi$ is a process locally in $\Lambda(\mathbb{L},\tilde{A},A,\lambda)$.

(ii) For every stopping time τ the random variable $a_\tau \xi$ depends only on the values of ξ on $[0,\tau[$.

(iii) For every $\beta > 0$, there exists an increasing adapted positive process L^β such that for every couple (ξ,ξ') of $\mathbb{H}$-valued regular processes for which $\sup_s \|\xi_s\|_{\mathbb{H}} \leq \beta$, and $\sup_s \|\xi'_s\| \leq \beta$, and for every $t < t' \in \mathbb{R}^+$ the following Lipschitz condition holds :

$$(L_1)\qquad \int_t^{t'} \lambda_s(\underset{\sim}{a}(\xi) - \underset{\sim}{a}(\xi'))\, dA_s < \int_t^{t'} \sup_{u<s} \|\xi_u - \xi'_u\|^2 \, dL_s^\beta$$

2.2 - Typical example

A typical example of equation (2.1) is the following :

$$\xi_t = V_t + \int_{]0,t]} \underset{\sim}{a}_s^1 \xi dS_s + \int_{]0,t]} \underset{\sim}{a}_s^2 \xi dM_s + \int_{]0,t]}\int_{\underline{E}} \underset{\sim}{a}^3(.,s,\xi,x) q(.,ds,dx)$$

where S is a $\mathbb{G}$-valued semi-martingale, M a $\mathbb{K}$-valued square integrable martingale ($\mathbb{G}$ and $\mathbb{K}$ Hilbert) and q is a white random measure of some order r. To S, M and F^q we associate the Λ-spaces defined in examples above and assume for the functionals $\underset{\sim}{a}^1$, $\underset{\sim}{a}^2$ and $\underset{\sim}{a}^3$ properties (i) a (iii) above.

By considering the process Z, with components S, M and F^q, taking its values in $\mathbb{G} \times \mathbb{K} \times M^r$ where M^r is the Banach space of measures weighted by $\frac{|x|^r}{|x|^r+1}$, we see immediately that the situation reduces to the one described in 2.1.

The reader will check for himself that the Lipschitz condition on $\underset{\sim}{a}^3$ expresses in our general context the one considered by A.V. Skorokhod (cf. [13]) and others ([4], [6]).

2.3 - Existence, uniqueness, non explosion theorems

Theorem 1. Under the assumptions made in 2.1 , there exists a unique stopping time τ and a unique (up to P-equivalence) process ξ on $[0,\tau[$ such that

(1) τ is predictable and on the set $\{\tau<\infty\}$ we have $\limsup_{t\uparrow\tau} \|\xi_t\|_{\mathbb{H}} = +\infty$ (i.e. when finite $\tau(\omega)$ is an explosion time).

(2) ξ is a strong solution of (2.1) on $[0,\tau[$.

Theorem 2. If, to the assumptions given in 2.1 , we add the following one : for every $t \in \mathbb{R}^+$

$$\int_0^t \lambda_s(\underset{\sim}{a}(\xi))\, dA_s \leq \int_0^t (1 + \sup_{u<s} \|\xi_u\|^2_{\mathbb{H}})\, dL_s$$

then $P\{\tau=\infty\} = 1$ (no explosion).

2.4 - Stability theorems

We consider two equations :

$$\xi_t = V_t + \int_{]0,t]} \underset{\sim}{a}\xi_s \, dZ_s \tag{2.2}$$

$$\xi'_t = V'_t + \int_{]0,t]} \underset{\sim}{a}'\xi'_s \, dZ'_s \tag{2.3}$$

of the type considered in 2.1.

We assume more precisely that Z and Z' are $\mathbb{B}$-valued regular processes controlled respectively in $\Lambda(\mathbb{L},\widetilde{A},A,\lambda)$ and $\Lambda(\mathbb{L},\widetilde{A}',A',\,')$ where $\mathbb{L}$ is a closed subspace of $\mathcal{L}(\mathbb{B}\,;\mathbb{H})$ and V and V' are regular $\mathbb{H}$-valued processes. The functionals $\underset{\sim}{a}$ and $\underset{\sim}{a}'$ verify the conditions (i) and (ii) with respect to the Λ-spaces considered for Z and Z' and the condition (iii) with the same Lipschitz coefficient-process L independent of $\beta > 0$.

The proximity of Z and Z' will be expressed through the consideration of $Z-Z'$ and making the following assumptions :

(iv) $Z-Z'$ is controlled in $\Lambda(\mathbb{L};\widetilde{Q},Q,\mu)$.

We assume moreover :

(v) for every regular processes ξ and ξ' and $t < t'$ $\mathbb{R}^+$

$$\int_t^{t'} \mu_s(\underset{\sim}{a}'(\xi) - \underset{\sim}{a}'(\xi'))\, dQ_s \leq \int_t^{t'} \sup_{u<s} \|\xi_u - \xi'_u\|^2_{\mathbb{H}}\, dL_s$$

ξ being the solution of (2.2) on $[0,\sigma[$ we define and assume :

(2.4) $d_1 := E\{\sup_{s<\sigma} \lambda_s(a\xi - a'\xi)\} < \infty$

(2.5) $d' := E\{\sup_{s<\sigma} \|V_s - V'_s\|^2_{\mathbb{H}}\} < \infty$

(2.6) $d_2 := E\{\sup_{s<\sigma} \mu_s(a\xi - a'\xi)\} < \infty$

(2.7) $c := E\{\sup_{s<\sigma} \mu^1_s(\underset{\sim}{a}\xi)\} < \infty$

Theorem 3. Under the above assumptions and the hypothesis that the positive random variables $A_{\sigma-}$, $\widetilde{A}_{\sigma-}$, $A'_{\sigma-}$, $\widetilde{A}'_{\sigma-}$, $Q_{\sigma-}$, $\widetilde{Q}_{\sigma-}$ are finite, the equation (2.3) has a (unique) strong solution ξ' on $[0,\sigma[$. Let $\xi > 0$ be given and q be a positive number such that $P\{\widetilde{Q}_{\sigma-} \vee \widetilde{Q}_{\sigma-} Q_{\sigma-} \geq q\} < \frac{\varepsilon}{2}$.

Then there exists a function $R_\varepsilon(d_1,d_2,q)$, determined by the functional $\underset{\sim}{a}$ and the processes $A, \widetilde{A}$ and L only, such that $\lim_{d_1,d_2,q\to 0} R_\varepsilon(d_1,d_2,q) = 0$, and a stopping time $\sigma_\varepsilon \leq \sigma$ such that :

(a) $P\{\sigma_\varepsilon < \sigma\} \leq \varepsilon$

(b) $E(\sup_{t<\sigma_\varepsilon} \|\xi_t - \xi'_t\|^2) \leq R_\varepsilon\,(d_1,d_2,q)$

Let $\ell > 0$ such that

$P\{\widetilde{A}_{\sigma-} A_\sigma \vee A'_\sigma - \vee L_{\sigma-} \geq \ell\} \leq \frac{\varepsilon}{2}$

A function R_ε is given by

$$R_\varepsilon(d_1,d_2,q) := 2K \sum_{j=0}^{[4\rho\ell]} [4\rho\ell]^j$$

where

$K := 6(d' + d_1\ell + d_2 q + cq)$, $\rho := \ell + q$

REFERENCES

1. C. Doléans-Dade. On the existence and unicity of solutions of stochastic integral equations. *Z.W.* 34 (1976), pp. 93-101.

2. C. Doléans-Dade, P.A. Meyer. Equations différentielles stochastiques. *Seminaire Prob. Strasbourg. XI.* Lectures Notes Math. 581, Springer Verlag.

3. M. Emery. Stabilité des solutions des équations différentielles stochastiques. Application aux intégrales multiplicatives stochastiques. *Z.W.* 41 (1978) pp. 241-262.

4. L. Galtchouk. Existence et unicité pour des équations différentielles stochastiques par rapport à des martingales et des mesures aléatoires. *2d Vilnius Conference Prob. Math. Stat.* 1 (1977), pp. 88-91.

5. K. Ito. On stochastic differential equations. *Mem. Amer. Math. Soc.* 4 (1951).

6. J. Jacod. *Calcul stochastique et problèmes de martingales.* Lecture Notes in Math. 714, Springer Verlag, 1979.

7. N. Kasamaki. On a stochastic integral equation with respect to a weak martingale. *Tôhoku Math. J.* 26 (1974) pp. 53-63.

8. M. Métivier.*Reelle und vektorwertige Quasimartingale und die Theorie der stochastischen Integration.* Lecture Notes Math. 602, Springer Verlag, 1977.

9. M. Métivier. *Stochastic equations driven by random measures and semimartingales.* To appear in *Contributions to probability.* J. Gani editor. Academic Press, 1981, pp. 173-188.

10. M. Métivier, J. Pellaumail. On a stopped Doob's inequality and general stochastic equations. *Ann. Prob.* Vol. 8, N° 1 (1980) pp. 96-114.

11. M. Métivier, J. Pellaumail. *Stochastic Integration.* Series of Monographies in Probability and Statistics. Academic Press, 1980.

12. M. Métivier, G. Pistone. Une formule d'isométrie pour l'intégrale stochastique Hilbertienne et équations d'évolution stochastique. *Z.W.* 33 (1975) pp. 1-18.

13. A.V. Skorokhod. *Studies in the theory of random processes.* Wiley, New York, 1965.

A MARTINGALE APPROACH TO FIRST PASSAGE PROBLEMS AND A NEW CONDITION FOR WALD'S IDENTITY

A.A. Novikov

Steklov Mathematical Institute
USSR, Moscow

Abstract. Some results about the distribution of passage times of processes with independent increments through non-linear boundaries are presented. The menthod for obtaining these results is based on the martingale technique. A new condition for Wald's identity is also presented.

1. Introduction. Let S_t be a process with independent homogeneous increments, $t \in R^+$ or $Z^+ = (0,1,\ldots)$, $S_o = 0$, $ES_1 = 0$, on a probability space $(\Omega, \mathcal{F}, P)$ with the natural filtration $\mathcal{F}_t = \sigma(S_u, u \le t)$. We consider the first passage time $\tau = \inf \{t \ge 0: S_t \notin D_t\}$, where D_t is a region with nonrandom boundaries (further $D_t = (-\infty, g(t))$, $g(0) > 0$, or $D_t = (f(t)- g(t), f(t) + g(t))$, $|f(0)| < g(0)$, $g(t) > 0$). In the case of D_t with linear boundaries there exists a great number of results about the distribution of τ and other related boundary functionals (see, f.e. Borovkov [1], Koroljuk [2], Skorokhod [3], Takács [4]).

The case of nonlinear boundaries is investigated much less extensively. Some results are obtained in this direction by Borovkov [5], Rotar' [6], Gut [7], Portnoy [8], Lai [9], Lai and Wijsman [10]. In the cited papers basically the direct probabilistic method was used. The application of the traditional methods of studying linear boundary problems (e.g. the method of factorization identities [1], the method of potentials [2], the method of differential equations [3], the combinatorial methods [4]) is connected with great difficulties. Note that some important results for the case of sums of independent random variables (that is the case of discret time in our interpretation) were obtained by Mogul'skij and Pecherskij [11], [12], using the technique of factorization identities.

The main tools of the martingale approach used in the present paper are based on Girsanov type theorems and on that well-known fact that for any uniformly integrable martingale Z_t and any stopping time τ the identity

$$(1) \qquad EZ_\tau = EZ_0$$

holds. (This is the famous Wald's indentity when $Z_t = S_t$ is a process with independent increments.) Choosing a suitable martingale Z_t it is possible to extract from this identity the useful information about the distribution of a stopping time τ. Of course, this identity was used many times in similar problems (see f.e. Shepp [13], Robbins and Siegmund [14], Novikov [15]).

Girsanov's type theorems can be applied in the following way. Let Z_t be a uniformly integrable positive martingale with $EZ_t = 1$. Then $\lim_{t\to\infty} Z_t = Z_\infty$ exists a.s. and $EZ_\infty = 1$. Introduce the new probability measure $\tilde{P}$ on $(\Omega, \mathcal{F})$, by

$$(2) \qquad \tilde{P}(A) = EI(A)\, Z_\infty,$$

where $I(A)$ is the indicator function. With respect to the new measure $\tilde{P}$ the process S_t will have some drift $a(t)$ and by a Girsanov type theorem the process $\tilde{S}_t = S_t - a(t)$ is a martingale. In some cases we can choose the martingale Z_t in such a way that the region D_t with nonlinear boundaries will be transformed (in some sense) in a region with linear boundaries for the martingale $\tilde{S}_t$ or for some other related processes. Assuming $A = (\tau > T)$ in (2) we can obtain the lower and upper bounds for the probability $P\{\tau > T\}$.

Now we are going to formulate some results obtained by the martingale method. Because of the limited space we give only sketches of the proofs. The full proofs will be published elsewhere.

2. The case of stable processes without positive jumps.

Consider the stable processes whose Laplace transform is

$$E\exp(\lambda S_t) = \exp(d\lambda^\alpha t) \quad (d > 0,\ \lambda \geq 0,\ 1 < \alpha \leq 2,\ t \in R^+).$$

It is known ([3]) that such stable processes have no positive jumps at all. For convenience of notation we assume $d = \frac{1}{\alpha}$ · (The general case can be easily obtained from here). Note that in the case $\alpha = 2$ S_t is the standard Wiener process.

Denote

$$H(\nu, \alpha, x) = \frac{1}{\Gamma(-\alpha\nu)} \int_0^\infty y^{-\alpha\nu-1} \exp(xy - \frac{1}{\alpha} y^\alpha)\, dy, \quad (\mathrm{Re}\nu < 0).$$

This function has analytical continuation to the region $\mathrm{Re}\nu \geq 0$ with poles on the line $\mathrm{Im}\lambda = 0$ (the first pole $\nu = 1$ when $1 < \alpha < 2$; the function $H(\nu, 2, x)$ has no poles at all). Let

$$\nu_\alpha(x) = \min(\nu > 0: H(\nu,\alpha,x) = 0).$$

It can be shown that $\nu_\alpha(x)$ is a monotonic continuous function such that $0 < \nu_\alpha(x) < 1/\alpha$ when $x > 0$, $\nu_\alpha(0) = \frac{1}{\alpha}$ and $\nu_\alpha(x) > 1/\alpha$ when $x < 0$.

In the next theorem we suppose $b^\nu H(\nu,\alpha,x) = c^{\alpha\nu}$ for $b = 0$ and $c > 0$.

<u>Theorem 1.</u> Let $E\exp(\lambda S_t) = \exp(\frac{1}{\alpha}\lambda^\alpha t)$, $\lambda \geq 0$, $1 < \alpha \leq 2$, $t \in R^+$ and

$$g(t) = a(t+b)^{1/\alpha} + c, \quad (b \geq 0, \quad ab^{1/\alpha} + c > 0).$$

If $\nu < \nu_\alpha(a)$ then $E\tau^\nu < \infty$ and

$$E(\tau + b)^\nu = b^\nu \frac{H(\nu,\alpha,-cb^{1/\alpha})}{H(\nu,\alpha,a)}. \tag{3}$$

If $\nu \geq \nu_\alpha(x)$ then $E\tau^\nu = \infty$.

The proof of this theorem is based on that fact that the process $\psi_t = (t+b)^\nu H(\nu,\alpha,(S_t - c)(t+b)^{-1/\alpha})$ is a martingale (when $1 < \alpha \leq 2$ and $\mathrm{Re}\,\nu < 1$ or when $\alpha = 2$ and any ν).

As $\psi_{t\wedge\tau}(\nu)$ is also a martingale hence

$$E\psi_{t\wedge\tau}(\nu) = \psi_0(\nu). \tag{4}$$

Using the fact that the excess S_t over the boundary $g(t)$ is zero, we can pass to the limit at $t \to \infty$ in (4) and obtain (3). (The details of the proof can be found in [16]).

3. <u>Asymptotic behaviour of</u> $P\{\tau > T\} = P\{S_t \leq g(t), 0 \leq t \leq T\}$ <u>when</u> $T \to \infty$.

The result of previous theorem seems to be exceptional in the sense that it hardly exists other class of processes with independent increments and non-linear boundaries $g(t)$, for which the moments of τ can be evaluated exactly. But it turned out that the asymptotic behaviour of $P\{\tau > T\}$ can be obtained under some general conditions on S_t and $g(t)$.

We have to suppose that the positive jumps of S_t satisfy the so-called right-side Cramer condition.

A_+: There exists a $\lambda > 0$ such that $E\exp(\lambda S_I) < \infty$.

In the case of discret time it will be convenient to define function g(t) for all $t \geq 0$ by means of the linear interpolation. We shall use the notations $f(t) << g(t)$ meaning that there exist some contants $C > 0$ and $t_o \geq 0$ such that $f(t) \leq C\, g(t)$ for all $t \geq t_o$.

Theorem 2. Let condition A_+ hold and set

$$(5) \qquad \ln E \exp(\lambda S_1) \equiv \varphi(\lambda) \sim d\lambda^{\alpha}, \qquad \lambda \downarrow 0,$$

where $d > 0$ and $1 < \alpha \leq 2$. Furthermore, suppose that: 1) $\sup g(t) < 0$, or 2) $g(t)$ is a continuous function, $0 << g(t) << t^{1/\alpha}$,

Then the expectation ES_τ exists, $ES_\tau \geq 0$ and

$$(6) \qquad P\{\tau > T\}T^{1/\alpha} \to ES_\tau (d^{1/\alpha}\Gamma(1-\tfrac{1}{\alpha}))^{-1}, \; T \to \infty.$$

The sketch of the proof. Using Wald's identity it is easy to show the existence of ES_τ and $ES_\tau \geq 0$. Then we show that $Z_{t\wedge\tau} = \exp(\lambda S_{t\wedge\tau} - \varphi(\lambda)t\wedge\tau)$ is a uniformly integrable martingale (for sufficiently small $\lambda > 0$) and hence identity

$$E \exp(\lambda S_\tau - \varphi(\lambda)\tau) = 1$$

holds. Using this fact we can show that

$$1 - E\exp(-\varphi(\lambda)\tau) \sim \lambda ES_\tau \quad \text{as} \quad \lambda \downarrow 0$$

and hence by the Tauberian theorem we obtain the asymptotic relation (6).

Remark 1. It follows from the Tauberian theorem that assumption (5) is fulfilled iff

$$P\{S_1 < x\} \sim \frac{d|x|^{-\alpha}}{|\Gamma(1-\alpha)|} \text{ as } x \to -\infty \text{ when } 1 < \alpha < 2$$

or $ES_1^2 < \infty$ when $\alpha = 2$.

If

$$P\{S_1 < x\} = \frac{dL(-x)|x|^{-\alpha}}{|\Gamma(1-\alpha)|}, \quad x \to -\infty, \quad 1<d<2,$$

where L(x) is a slowly varying function then the conjunction (6) remains true but with function $\gamma(T)$ instead $T^{1/\alpha}$, where $\gamma(T)$ is the inverse function of $T^{\alpha}/L(T)$.

Remark 2. Theorem 2 gives the exact asymptotics for $P(\tau > T)$ only if $0 < ES_\tau < \infty$. It is easy to show that under conditions of Theorem 2 $E|S_\tau| < \infty$ and

(7) $$\int_1^\infty \frac{|g(t)|}{t^{1+1/\alpha}} dt = \infty \Rightarrow ES_\tau = 0,$$

hold true when g(t) is a continuous non-increasing function, while $ES_\tau > 0$ and

(8) $$\int_1^\infty \frac{g(t)}{t^{1+1/\alpha}} dt = \infty \Rightarrow ES_\tau = \infty,$$

hold when g(t) is a non-decreasing function $(g(1)>0,\ t\in Z^+)$.

The problem of finding necessary and sufficient conditions for inequalities $0 < ES_\tau < \infty$ appeared to be nontrivial. We resolved this problem here under some additional conditions on jumps S_t and the boundary g(t). We have to suppose that the positive jumps S_t are bounded and that the negative jumps S_t satisfy the left-side Cramer condition.

A_-: There exist a $\lambda < 0$ such that $E\exp(\lambda S_1) < \infty$. Of course, under these conditions parameter α in (5) is two and all moments of S_t exist.

Denote

$\lambda_- = \inf(\lambda < 0\colon E\exp(\lambda S_1) < \infty)$.

Theorem 3. Let the positive jumps of S_t, $t\in R^+$, be bounded by some constant and let condition A_- hold.

1°) If g(t) is a convex nonincreasing function there exists a number $t_o > 0$ such that $g(t_o) > 0$ and $g'(t_o) > \varphi'(\lambda_-)$ then $E|S_\tau| < \infty$ and

(9) $$\int_1^\infty \frac{|g(t)|}{t^{3/2}}\, dt < \infty \iff ES_\tau = 0.$$

2°) If g(t) is a concave non-decreasing function then $ES_\tau > 0$ and

(10) $$\int_1^\infty \frac{g(t)}{t^{3/2}}\, dt < \infty \iff ES_\tau < \infty.$$

The sketch of the proof. Let t_o be a number as indicated in 1°) or such that $g'(t_o) < \psi'(+\infty)$ in the case 2°). Define b(t) as the solution of the equation

$$\psi'(b(t)) = g'(t), \qquad t > t_o,$$

and assume $b(t) = 0$ at $0 \le t \le t_o$. Then it is easy to check that the process

$$Z_t = \exp\{\int_o^t b(u)\, dS_u - \int_o^t \psi(b(u))du\}$$

is a uniformly integrable martingale and hence $EZ_\infty = 1$. Now introduce the new probability measure $\tilde{P}$ by the formula (2) and denote the integration by this measure by the symbol $\underline{M}$. With respect to the measure $\tilde{P}$ the process

$$\tilde{S}_t = S_t - \int_o^t \psi'(b(u))du, \quad t \ge 0,$$

is a martingale. Using this fact we can show that

$$T^{-1/2} << \tilde{P}\{\tau > T\} << T^{-1/2}.$$

Then from the equality

$$P\{\tau > T\} = MI\{\tau > T\}\, Z_T^{-1}$$

we can deduce (using the Jensen's inequality and some other arguments) that under the condition $\int_1^\infty |g(t)|t^{-3/2} dt < \infty$ the following relation holds

(11) $$T^{-1/2} << P\{\tau > T\} << T^{-1/2}.$$

From here and from (6) the forward implications in (9) and (10) follow. The backward implications in (9) and (10) were given earlier under less restrictive conditions (see (7) and (8)).

The proof of Theorem 3 and partly of Theorem 2 will be given in [16]. The case of boundaries g(t) such that $-t \ll g(t) \ll -t^{-1/\alpha}$ was considered in [17] (see also [16]; in this case only the rough asymptotics of $P\{\tau > T\}$ was found.)

Note that studying some nonlinear parabolic equations by probabilistic methods Gertner [18] and Uchiyma [19] obtained independently the inequalities of (11) for the case of the Wiener process.

4. A condition for Wald's identity.

Here we give a sufficient condition on a stopping time σ (not obligatorily a first passage time) which ensures the validity of Wald's identity $ES_\sigma = 0$. Recall that we consider the case of the processes with independent homogeneous increments and $ES_1 = 0$.

Let $\mathfrak{N}$ denote the class of positive continuous nondecreasing and convex (for all sufficiently large t) functions g(t) such that

$$\int_1^\infty \frac{g(t)}{t^{3/2}}\, dt = \infty$$

(e.g. $g(t) = \sqrt{t}$ or $\sqrt{t}(\log(t+2))^{-1}$ etc.).

Theorem 4. Let conditions A_+ and A_- hold and let $E|S_\sigma| < \infty$. If $g(t) \in \mathfrak{N}$ then

$$Eg(\sigma) < \infty \Rightarrow ES_\sigma = 0.$$

The sketch of the proof. Introduce the auxilary stopping time

$$\tau_a = \inf\{t \geq 0 : S_t > a - g(t)\}, \quad a > g(0).$$

From the implication (7) follows that $ES_{\tau_a} = 0$ and hence

$$ES_{\tau_a \wedge \sigma} = EI(\tau_a \leq \sigma) S_{\tau_a} + EI(\tau_a > \sigma)\, S_\sigma = 0.$$

Obviously, $\tau_a \to \infty$ a.s. when $a \to \infty$. As $E|S_\sigma| < \infty$

$$EI\{\tau_a > \sigma\} S_\sigma \to ES_\sigma, \ (a \to \infty).$$

On the other hand

$$EI(\tau_a \le \sigma)\, S_{\tau_a} \le EI(\tau_a \le \sigma)\,(a - g(\tau_a)) \ge$$

$$\ge - EI(\tau_a \le \sigma)\, g(\sigma) \to 0, \quad a \to \infty,$$

in view the assumption $Eg(\tau) < \infty$. So we have $ES_\sigma \le 0$. Using the stopping time $\tau'_a = \inf\{t \ge 0: S_t \ge g(t) - a\}$ just in the same manner we can show $ES_\sigma \ge 0$ and hence Theorem 4 holds.

Remark 3. It is interesting to compare Theorem 4 with a recent result of Azema, Gundy and Yor [20] about conditions for the uniform integrablity of continuous martingales. In particular from [20] it follows that if

(12) S_t is the Wiener process and $\sup\limits_{T \ge 0} E|S_{T\wedge\sigma}| < \infty$

then

(13) $\lim\limits_{T\to\infty} P\{\sigma > T\} T^{1/2} = 0 \Leftrightarrow S_{\sigma\wedge T}$ is uniformly integrable $\Rightarrow ES_\sigma = 0$

It is easy to check that if $P\{\sigma > T\}\, T^{1/2} \to 0$, $T \to \infty$, then there exists a function $g(t) \in \mathfrak{N}$ such that $Eg(\sigma) < \infty$ and hence by Theorem 4

$$\lim_{T\to\infty} P\{\sigma > T\}\, T^{1/2} = 0 \Rightarrow ES_\sigma = 0$$

when

(14) S_t satisfies A_+ and A_- and $E|S_\sigma| < \infty$.

So we obtained the same (as in (13)) condition for Wald's identity under a less restrictive assumption ((14) instead of (12)), but only for a special type of martingales (with independent, homogeneous increments).

5. Asymptotics of $P\{\tau > T\} = P\{|S_t + f(t)| \le g(t),\ 0 \le t \le T\}$ when $T \to \infty$.

The case of two-sided boundaries is more difficult and we received only a rough asymptotics for some classes of processes S_t.

<u>Theorem 5.</u> Let S_t be a Gaussian process, $t \in R^+$ or $t \in Z^+$ and

$$\tau = \inf \{t: |S_t + f(t)| \geq g(t)\},$$

where:

1) $g(t)$ is a positive function, $g(t) \to \infty$ as $t \to \infty$, $\int_0^\infty g'^2(t)\,dt < \infty$;

2) $f(t)$ is a continuous function, $f'(t) \downarrow 0$ as $t \to \infty$. Then

$$1 + \int_0^T (f'(t))^2\,dt = o\left(\int_0^T g^{-2}(t)\,dt\right) \Rightarrow \ln P\{\tau > T\} \sim -\frac{\pi^2}{8} \int_0^T g^{-2}(t)\,dt,$$
$$T \to \infty$$

and

$$\int_0^T g^{-2}(t)\,dt = o\left(\int_0^T (f'(t))^2\,dt\right) \Rightarrow \ln P\{\tau > T\} \sim -\frac{1}{2} \int_0^T f'^2(t)\,dt,$$
$$T \to \infty.$$

Under some mild additional assumptions on functions $f(t)$ and $g(t)$ the result of Theorem 5 remains true for a non-gaussian $S_t (t \in Z^+)$ with bounded jumps as well (see [21]). Results of another type were obtained by Lai and Wijsman [10]. The case $f(t) = 0$ was considered also in [17] and [22].

REFERENCES

[1] A.A.Borovkov, Stochastic Processes in Queuing Theory, Nauka, Moscow, 1972. (English translation: Springer-Verlag, 1976).

[2] V.S.Koroljuk, Bondary problems for Compound Poisson Processes, Naukova Dumka, Kiev, 1975.

[3] A.V.Skorokhod, Random Processes with Independent Increments, Nauka, Moscow, 1964.

[4] L.Takács, Combinatorial Methods in the Theory of Stochastic Processes, New York, 1967.

[5] A.A.Borovkov, "Boundary problems for random walks and large deviations in function spaces", Theory Prob. Applications, vol. 12, No. 3, 1967.

[6] V.I.Rotar', "On moments of the time and the value of the first jump over a curvilinear bound", Theory Prob.Applications, vol. 12, No. 3, 1967.

[7] A.Gut, "On the moments and limit distributions of some passage time, Annals Prob., vol. 2, No. 2, 1974.

[8] St.Portnoy, "Probability bounds for the first exit through moving boundaries", Annals Prob., vol. 6, No. 1, 1978.

[9] T.L.Lai, "First exit times from moving boundaries for sums of independent random variables", Annals Prob., vol. 5, No. 2, 1977.

[10] T.L.Lai, R.A.Wijsman, "First exit time of a random walk from the bounds $f(n) \pm g(n)$, with applications", Annals Prob., vo. 7, No. 4, 1979.

[11] A.A.Mogul'skij, E.A.Pecherskij, "On a first exit time of a random walk from a semigroup in R^n", Theory Prob. Application, vol. 22, No. 4, 1977.

[12] A.A.Mogul'skij, E.A.Pecherskij, "On a time of the first hit in the region with a curved bound", Sibirian Math. I., vol. 19, No. 4., 1978.

[13] L.A.Shepp, "A first passage problem for the Wiener process", Ann. Math. Stat., vol. 38, No. 6, 1967.

[14] H.Robbins, D.Siegmund, "Boundary crossing probabilities for the Wiener process and sample sums", Ann. Math. Stat., vol. 41, No. 5, 1970.

[15] A.A.Novikov, "On stopping times for the Wiener process", Theory Prob. Applications, vol. 16, No. 3., 1971.

[16] A.A.Novikov, "Martingale approach to first passage problems for nonlinear boundaries", Proceedings of the Steklov Inst. of Math., vol. 158, 1981.

[17] A.A.Novikov, "On bounds and asymptotic behavior of moving boundary crossing probabilities for sums of independent random variables", Transaction of Academy of Science of USSR, Seria Math., vol. 44, No. 4, 1980.

[18] I.Gertner, "Upper and lower bounds for brownian first exit densities and propagation of wave fronts", preprint, 1980.

[19] K.Uchiyma, "Bronwian first exit form and sojourn over one-sided moving boundary and application", preprint, 1980.

[20] I.Azéma, R.Gundy, M.Yor, "Sur l'integrabilité uniforme des martingales continues, Lecture Notes in Math., vol. 784, 1980.

[21] A.A.Novikov, "On a first exit time of sums of bounded random variables from a curvilinear strip", Theory Prob. Applications, vol. 26., No. 3, 1981.

[22] A.A.Novikov, "On bounds and asymptotic behavior of moving boundary noncrossing probabilities for the Wiener process", Matem. Sbornik, vol. 110 (152), No. 4 (12), 1979.

Steklov Mathematical Institute
Vavilov str. 42, Moscow,
117966 GSP-1, USSR

A TAYLOR FORMULA FOR SEMIMARTINGALES SOLVING A STOCHASTIC EQUATION

Eckhard Platen
Akademie der Wissenschaften der DDR
Institut für Mathematik
Berlin

Abstract

The paper presents a generalized Taylor formula for solutions of stochastic equations which are semimartingales.

1. Introduction

By the use of the Ito formula for semimartingales of Galtchouk [1] we formulate in the following a Taylor formula which generalizes our earlier results in [3] and [4]. This result is very general and allows the separate consideration of predictable jumps of the semimartingale. For instance, the Taylor formula is useful for the construction of time discrete approximations of solutions of stochastic equations (see Platen [3]). But it may be also a helpful tool in other theoretical and practical investigations as the well-known Taylor formula in the non-stochastic case.

2. Stochastic Equation

At first some notational conventions. For unexplained notations, terminology or assertions we refer to Jacod [2].

Let $(\Omega,\underline{\underline{F}},P)$ be a complete probability space and $\underline{\underline{F}} = (\underline{\underline{F}}_t)_{t\in R_+}$ an increasing right-continuous family of complete sub-σ-fields of $\underline{\underline{F}}$. $\underline{\underline{P}}$ (resp. $\underline{\underline{O}}$) denotes the predictable (resp. optional) σ-field of $\Omega \times R_+$ generated by all continuous (resp. right-continuous left-hand limited) adapted processes. The progressive σ-field $\underline{\underline{\Pi}}$ is the class of all subsets $A\subset \Omega \times R_+$ such that $A \cap (\Omega \times [o,t])$ is $\underline{\underline{F}}_t \otimes \underline{\underline{B}}(R_+)$-measurable for each $t\in R_+$.

$\underline{\underline{E}}$ denotes the Borel-$\mathfrak{b}$-field on $E = R^d \setminus \{0\}$, $d \in N = \{1,2,\ldots\}$, and we set $\tilde{\underline{\underline{P}}} = \underline{\underline{P}} \otimes \underline{\underline{E}}$, $\tilde{\underline{\underline{O}}} = \underline{\underline{O}} \otimes \underline{\underline{E}}$. $\underline{\underline{V}}$ denotes the set of adapted processes with finite variation on R_+. $\underline{\underline{A}}$ is the set of processes A of $\underline{\underline{V}}$ with integrable variation on R_+, which means $E \int_0^\infty |d A_s| < \infty$.

If $\underline{\underline{J}}$ is a set of processes, then we denote by $\underline{\underline{J}}_{loc}$ the set of processes $Y = (Y_t)$ for which a sequence of stopping times (T_n), $n \in N$, exists which is increasing to $+\infty$ and $Y_{t \wedge T_n} \in \underline{\underline{J}}$ for each $n \in N$.

In the following we consider the n-dimensional semimartingale $X = (X^j)_{j \le n}$, right-continuous with left-hand limits, which fulfils the stochastic equation

(1)
$$X_t = X_o + \sum_i \{(a^i(X_-) \cdot A^i)_t + (b^i(X_-) \cdot M^i)_t + (c^i(X_-) I_{|c| \le 1} \cdot (\mu - \nu))_t + (d^i(X_-) I_{|c| > 1} \cdot \mu)_t + (e^i(X_-) \cdot p)_t\},$$

where X_o is $\underline{\underline{F}}_o$-measurable and $a^i(X_-) \in \underline{\underline{G}}_{(i)}$, $b^i(X_-) \in \underline{\underline{G}}_{(d+i)}$, $c^i(X_-) I_{|c| \le 1} \in \underline{\underline{G}}_{(2d+i)}$, $d^i(X_-) I_{|c| > 1} \in \underline{\underline{G}}_{(3d+i)}$ and $e^i(X_-) \in \underline{\underline{G}}_{(4d+i)}$, $i \in D = \{1,\ldots,d\}$. $a^{ji}, b^{ji}, \ldots, e^{ji}$ are real-valued functions on R^n or $R^n \times E$, resp. We write X_- for the process (X_{t-}) and $|c| = \sum_{i,j} c^{ji}(X_-)$. I_G is the indicator of the set G.

$A^i \in \underline{\underline{A}}_{loc}$, $i \in D$, denotes a $\underline{\underline{P}}$- measurable one-dimensional continuous process starting form 0. By $\underline{\underline{G}}_{(i)}$, $i \in D$, we denote the set of $\underline{\underline{P}}$-measurable n-dimensional processes $H = (H^j)_{j \le n}$ for which

$$\int_0^t |\sum_j H_s^j| \, d A_s^i < \infty$$

P - a.s. for each $t \in R_+$. For such an H we put

$$(H \cdot A^i)_t = \int_0^t H_s \, d A_s^i = (\int_0^t H_s^j \, d A_s^i)_{j \le n} ,$$

where the integrals with respect to A^i are defined as usual Stieltjes integrals for each $\omega \in \Omega$. M^i, $i \in D$, is a one-dimensional continuous local martingale starting from 0. There exists an increasing continuous process

C with $C_0 = 0$ and $C_t < \infty$ for each $t \in R_+$ and a $\underline{\underline{P}}$-measurable process $\overline{q} = (\overline{q}^{ij})_{i,j\in D}$ with values in the set of symmetric nonnegative d×d-matrices, such that

$$(\overline{q}^{ij} \cdot C)_t = \int_0^t \overline{q}_s^{ij} \, d\, C_s = \langle M^i, M^j \rangle_t .$$

$\langle M^i, M^j \rangle_t$ is the only one $\underline{\underline{P}}$-measurable process with finite variation on each finite interval such taht $M^i M^j - \langle M^i, M^j \rangle$ is a local martingale starting from 0.

By $\underline{\underline{G}}_{(d+i)}$, $i\in D$, we denote the set of all $\underline{\underline{\Pi}}$-measurable n-dimensional processes $K = (K^j)_{j\le n}$ for which

$$(|K^j|^2 \, \overline{q}^{ii} \cdot C)_t < \infty$$

P - a.s. for all $t \in R_+$. For such a K we define

$$(K \cdot M^i)_t = \int_0^t K_s \, d\, M_s^i$$

as the only n-dimensional continuous local martingale which satisfies

$$\langle (K^j \cdot M^i), Y \rangle_t = (K^j \overline{b}^i \cdot C)_t$$

for each $j \le n$ and any one-dimensional continuous local martingale Y, where the process $\overline{b}^i$ is such that

$$\langle M^i, Y \rangle_t = (\overline{b}^i \cdot C)_t .$$

We remark that for $K \in \underline{\underline{G}}_{(d+i)}$ and $H \in \underline{\underline{G}}_{(d+j)}$, $i,j\in D$:

$$\langle (K^r \cdot M^i), (H^k \cdot M^j) \rangle_t = (K^r H^k \overline{q}^{ij} \cdot C)_t$$

for all $r,k \le n$.

$\underline{\underline{G}}_{(5d+2)}$ denotes the set of $\underline{\underline{\Pi}}$-measurable n-dimensional processes $K = (K^j)_{j\le n}$ for which

$$((\sum_j |K^j|^2)^{1/2} \cdot C)_t < \infty$$

P - a.s. for each $t \in R_+$.

Let Z be a d-dimensional semimartingale, right-continuous with left-hand limits, which is the sum of a local martingale and a process with finite variation on each finite interval. A sequence of totally inaccessible stopping times $(T_n), n \in N$, and a sequence of predictable stopping times $(S_n), n \in N$, exist which absorbs all discontinuity times of Z, that means all times $\hat{T}$ where $\Delta Z_{\hat{T}} = Z_{\hat{T}} - Z_{\hat{T}-} \neq 0$. The graphs of all S and T are disjoint.

On sets $G \in \underline{\underline{B}}(R_+) \times E$ we define the integer-valued measures

$$\mu(G) = \sum_n I_G(T_n, \Delta Z_{Tn})$$

and

$$p(G) = \sum_n I_G(S_n \Delta Z_{S_n}).$$

ν (resp. λ) is the dual predictable projection of μ(resp. p) and is defined as a positive random measure on $(R_+ \times E, \underline{\underline{B}}(R_+) \otimes \underline{\underline{E}})$ such that for any $G \in \tilde{\underline{\underline{P}}}$ $I_G \cdot \nu$ is $\underline{\underline{P}}$-measurable and

$$E(I_G \cdot \nu)_t = E(I_G \cdot \mu)_t$$

for each $t \in R_+$.

We denote by $\underline{\underline{G}}_{(2d+1)} = \underline{\underline{H}}'_{loc}(\mu)$, $i \in D$, the set of $\tilde{\underline{\underline{P}}}$-measurable n-dimensional functions $V = (V^j)_{j \leq n}$ on $\Omega \times R_+ \times E$ such that

$$(|V|^2 I_{|c| \leq 1} \cdot \mu) \in \underline{\underline{A}}_{loc}.$$

For such a V it is defined the n-dimensional stochastic integral

$$N_t = (VI_{|c| \leq 1} \cdot (\mu - \nu))_t$$

as the only purely discontinuous local martingale such that

$$\Delta N_t = \sum_n V(T_n, \Delta Z_{T_n}) I_{t = T_n} I_{|c| \leq 1}$$

P - a.s. for each $t \in R_+$.

$\underline{\underline{G}}_{(5d+1)}$ is the set of $\tilde{\underline{\underline{P}}}$-measurable n-dimensional functions $V = (V^j)_{j \leq n}$ on $\Omega \times R_+ \times E$ such that

$$(|V|^2 I_{|c| \leq 1} \cdot \nu)_t < \infty$$

P - a.s. for each $t \in R_+$.

We denote by $\underline{\underline{H}}''(\mu)$ (resp. $\underline{\underline{H}}''(p)$) the set of $\tilde{\underline{\underline{O}}}$-measurable n-dimensional functions $V = (V^j)_{j \leq n}$ on $\Omega \times R_+ \times E$ such that

$$(|V|I_{|c|>1} \cdot \mu)_{\cdot} = \sum_{T_n \le \cdot} |V(T_n, \Delta Z_{T_n})| I_{|c|>1} \in \underline{\underline{V}}$$

(resp. $(|V|\cdot p)_{\cdot} = \sum_{S_n \le \cdot} |V(S_n, \Delta Z_{S_n})| \in \underline{\underline{V}}$

P - a.s. and set $\underline{\underline{G}}_{(3d+i)} = \underline{\underline{H}}''_{loc}$, $\underline{\underline{G}}_{(4d+i)} = \underline{\underline{H}}''_{loc}(p)$, $i \in D$.

3. Itô Formula

$\underline{\underline{H}}'(p)$ denotes the set of $\tilde{\underline{\underline{P}}}$-measurable functions g with

$$E\,((|g|^2 \cdot p)_\infty)^{\frac{1}{2}} < \infty$$

and for each predictable stopping time S it is

$$E\,(g(S, \Delta Z_S) \mid \underline{\underline{F}}_{S-}) = 0$$

P - a.s. on the set $(S < \infty)$. Further it is

$$\underline{\underline{H}}(p) = \{g : g = h' + h'',\ h' \in \underline{\underline{H}}(p),\ h'' \in \underline{\underline{H}}''(p)\}.$$

We assume for all $g \in \underline{\underline{H}}'(\mu)$, $\underline{\underline{H}}'(p)$, $\underline{\underline{H}}''(\mu)$, $\underline{\underline{H}}''(p)$, $\underline{\underline{H}}(p)$ $g(\omega,t,o) \equiv 0$.

Further, let us assume for all $i \in D$:

$c^i(X_-)I_{|c| \le 1}$ and $d^i(X_-)\, I_{|c| > 1}$ are $\tilde{\underline{\underline{P}}}$-measurable,

$d^i(X_-)I_{|c| > 1} \in \underline{\underline{H}}''(\mu)$, $e^i(X_-) \in \underline{\underline{H}}(p)$ and

$(|c^i(X_-)|^2 I_{|c| \le 1} \cdot \mu) \in \underline{\underline{A}}_{loc}$.

To formulate the Itô formula we need some further notations.

Let

$$Q = \{(j_1,\dots,j_k) : k \in N,\ j_i \in \{1,\dots,5d+2\} \text{ for } i \in \{1,\dots,k\}\} \cup \{v\}$$

denote the set of row vectors $\alpha = (j_1,\dots,j_k)$ with finite length $l(\alpha) = k$, where $l(v) = 0$. We write $-\alpha$ or $\alpha-$ if we delete the first or last component of $\alpha \in Q$, $l(\alpha) \ge 1$, resp. By $s(\alpha)$ we denote the number of components of $\alpha \in Q$ which are elements of $\{2d+1,\dots,5d,\ 5d+1\}$.

For $\alpha = (j_1,\dots,j_k) \in Q$, $t \in R_+$ and functions $g \in \underline{\underline{G}}_\alpha$ we define recoursively the following multiple stochastic integrals

$$H_\alpha(g(\cdot))_t = \begin{cases} g(t) & \text{for } k = 0 \\ (H_{\alpha-}(g(\cdot))_- \cdot A^{j_k})_t & \text{for } k\ge 1,\quad 1\le j_k\le d \\ (H_{\alpha-}(g(\cdot))_- \cdot M^{j_k})_t & \text{for } k\ge 1,\quad d+1\le j_k\le 2d \\ (H_{\alpha-}(g(\cdot))_- \cdot I_{|c|\le 1}\cdot(\mu-\nu))_t & \text{for } k\ge 1,\quad 2d+1\le j_k\le 3d \\ (H_{\alpha-}(g(\cdot))_- * I_{|c|>1}\cdot\mu)_t & \text{for } k\ge 1,\quad 3d+1\le j_k\le 4d \\ (H_{\alpha-}(g(\cdot))_- \cdot p)_t & \text{for } k\ge 1,\quad 4d+1\le j_k\le 5d \\ (H_{\alpha-}(g(\cdot))_- \cdot I_{|c|\le 1}\cdot\nu)_t & \text{for } k\ge 1,\quad j_k = 5d+1 \\ (H_{\alpha-}(g(\cdot))_- \cdot C)_t & \text{for } k\ge 1,\quad j_k = 5d+2 \end{cases}$$

For $k \ge 2$ $\underline{\underline{G}}_\alpha$ is defined as the set of functions $g|\Omega \times R_+ \times E^{s(\alpha)} \to R^n$ for which

$$g\in\underline{\underline{G}}_{(j_1)},\ H_{(j_1)}(g(\cdot))_- \in \underline{\underline{G}}_{(j_2)},\ldots,\ H_{\alpha-}(g(\cdot))_- \in \underline{\underline{G}}_{(j_k)}.$$

Let C^2 denote the set of functions $F/R^n \to R^n$ with partial derivatives

$$D^i F = \frac{\partial}{\partial x_i} F \text{ and } D^{ir} F = \frac{\partial^2}{\partial x^i \partial x^r} F \text{ for all } i,r \in\{1,\ldots,n\}.$$

We introduce now the following operators on C^2:

$$L^i F(Y_t) = \sum_j a^{ji}(Y_{t-})\, D^j F(Y_{t-}),$$

$$L^{d+i} F(Y_t) = \sum_j b^{ji}(Y_{t-})\, D^j F(Y_{t-}),$$

$$L^{2d+i} F(Y_t) = F_{c^i}(Y_{t-}),$$

$$L^{3d+i} F(Y_t) = F_{d^i}(Y_{t-}),$$

$$L^{4d+i} F(Y_t) = F_{e^i}(Y_{t-}),$$

for $i \in\{1,\ldots,d\}$ and

$$L^{5d+1} F(Y_t) = \sum_i \{F_{c^i}(Y_{t-}) - \sum_j c^{ji}(Y_{t-})\, D^j F(Y_{t-})\}$$

$$L^{5d+2} F(Y_t) = \frac{1}{2} \sum_{r,k} D^{rk} F(Y_{t-}) \sum_{i,\ell} b^{ri}(Y_{t-})\bar{c}_t^{i\ell}\, b^{k\ell}(Y_{t-}),$$

where Y_t is a right-continuous left-hand limited process and

$$F_g(Y_{t-}) = F(Y_{t-} + g(Y_{t-})) - F(Y_{t-}).$$

Now we can write down the Itô formula of Galtchouk [1] for $F \in C^2$ and $t \in R_+$ in the form

$$(2) \qquad F(X_t) = F(X_o) + \sum_{r=1}^{5d+2} H_{(r)} (L^r F(X.))_t .$$

4. Taylor Formula

For all $\alpha = (j_1,\dots,j_k) \in Q$ we define the coefficient function $F_\alpha | R^n \times E^{s(\alpha)} \to R^n$ by

$$F_\alpha (Y_t) = \begin{cases} o & \text{for } k = 0,\ \alpha = V \\ a^{j_1}(Y_{t-}) & \text{for } k = 1,\ 1 \le j_1 \le d \\ b^{j_1}(Y_{t-}) & \text{for } k = 1,\ d+1 \le j_1 \le 2d \\ c^{j_1}(Y_{t-}) & \text{for } k = 1,\ 2d+1 \le j_1 \le 3d \\ d^{j_1}(Y_{t-}) & \text{for } k = 1,\ 3d+1 \le j_1 \le 4d \\ e^{j_1}(Y_{t-}) & \text{for } k = 1,\ 4d+1 \le j_1 \le 5d \\ o & \text{for } k = 1,\ 5d+1 \le j_1 \le 5d+2 \\ L^{j_1} F_{-\alpha} (Y_t) & \text{for } k \ge 2,\ 1 \le j_1 \le 5d+2, \end{cases}$$

where Y_t is a right-continuous left-hand limited process.

If we set for $U \subset Q$

$$B(U) = \{\alpha \in Q \backslash U : \ -\alpha \in U\},$$

then we can formulate the Taylor formula for the solution of the stochastic equation (1):

Theorem:

If for $U \subset Q$:

(i) $U \neq \emptyset$ and $\sup_{\alpha \in U} l(\alpha) < \infty$,

(ii) for all $\alpha \in U \setminus \{v\}$: $-\alpha \in U$,

(iii) for all $\alpha \in (U \cup B(U)) \setminus \{v\}$: $F_{-\alpha} \in C^2$ and $F_\alpha(X) \in \underline{\underline{G}}_\alpha$

(iv) for all $\alpha \in U$: $\tilde{F}_\alpha \equiv F_\alpha(X_o) \in \underline{\underline{G}}_\alpha$,

then for all $t \in R_+$:

$$X_t = X_o + \sum_{\alpha \in U} H_\alpha(F_\alpha(X_o))_t + \sum_{\alpha \in B(U)} H_\alpha(F_\alpha(X.))_t.$$

The proof of the theorem is using an iterated application of the Itô formula (2) and is formally the same as in Platen [3] or [4].

References

[1] Galtchouk, L.I.: On the predictable jumps of martingales. Proceedings of the Conference on Stochastic Differential Systems held in Vilnius 1978. Lect. Notes in Control and Inf. Sciences 25, Springer (1980), 50-57.

[2] Jacod, J.: Calcul stochastique et problemes de martingales. Lect. Notes in Math. 714, Springer (1979).

[3] Platen, E.: An approximation method for a class of Ito processes Liet. matem. rink. (1981) (to appear).

[4] Platen, E.: A generalized Taylor formula for solutions of stochastic equation. SANKHYA, Ser. A (1981) (to appear).

Akademie der Wissenschaften der DDR
Institut für Mathematik
DDR-1080 Berlin,
Mohrenstrasse 39

ON OPTIMAL SENSOR LOCATION IN STOCHASTIC DIFFERENTIAL SYSTEMS AND IN THEIR DETERMINISTIC ANALOGUES

Gy. Sonnevend
Dept. of Numerical Math.
Eötvös University
Budapest, HUNGARY

Introduction. We study here the problem of best choice of a fixed number, N, of linear measurements (nodes, sensors) for the L_2 approximation of functions with bounded energy defined over a bounded domain, G, in R^k and satisfying a given set of linear homogeneous bundary conditions. These problems (of "restricted N-width") are shown to be closely related to problems of optimal filtering ("smoothing") of a class of Gaussian random fields defined over G and specified by the Green function, corresponding to the energy space, i.e. to a 2m order strongly elliptic differential operator and the given boundary conditions, as the covariance function.

After proving explicit formulas and "duality" between minimum norm extremals (splines) and optimal filters (for fixed choices of nodes), following a line of works by Kalman, Ciesielski and others, we continue by deriving exact characterizations of the optimal sets of N nodes and exact lower bounds for the corresponding errors, showing the role of the zero sets of eigenfunctions. It turns out that in the deterministic case the best systems of N nodes are, in general, not unique for k > 1, while in the stochastic case the first N eigenfunctions are the unique best linear sensors.

Preliminaries. Although the following optimal approximation problems could be formulated in a more general setting,(see [10]), we restrict ourselves to the given class for simplicity and in order to be able to give explicit expressions for the optimal approximation operators and their errors (through eigenfunctions and Green functions of differential operators). For the used facts concerning the spaces, operators, processes introduced in this section we refer to [4] and [5]. - Let G be a bounded domain in R^k with piecewise C^{m-1} boundary, g, 2m > k, and X be the Hilbert space obtained from the scalar product

(1) $\langle u,v\rangle := \int_G \sum_{|i|,|j|\le m} a_{ij}(t)\, D^i u(t)\, D^j r(t), \quad a_{ij} = a_{ji}$

- corresponding to a quadratic (energy) expression - for functions u,v satisfying a given set of boundary conditions

(2) $$0 = B_i(x) = \sum_{|j| \le m-1} b_{ij}(t)\, D^j x(t),\ t \in g,\ i = 1,\ldots, r > 0.$$

The conditions (2), together with

(3) $$||x||^2 = \langle x,x \rangle \le 1$$

are assumed to imply that

$$(x,x)^2 = ||x||_o^2 = ||x||_{L^2(G)} \le c(a_{ij}, G, B_i) < \infty,$$

then the condition of strong ellipticity

$$\left| \mathrm{Re} \sum_{|i|,|j| = m} a_{ij}(t)\ z^i z^j \right| \ge C ||z||^{2p}$$

implies, by Gårding's inequality and Sobolew Lemma that the linear functionals

(4) $$f(x) = f_t(x) = x(t),\ t \in G,\ (2m > k),$$

are continuous over X, the functions x have $(m-{}^k/_2)$ continuous derivatives on G, and the embedding $X \to L_2(G)$ is a compact, symmetric (Hilbert-Schmidt) operator, $C^{1/2}$

(5) $$\langle Cx,x \rangle = (x,x).$$

There exists a natural self adjoint extension of the operator

$$A = \sum_{|i|,|j| \le m} (-1)^{|j|} D^j (c_{ij}(t)\ D^i),\ D^i = \frac{\partial^i x}{\partial t_1^{i_1} \ldots \partial t_k^{i_k}},\ |i| = \sum_{j \le k} i_j$$

defined on the subspace, X^o, of x specified by the so called natural boundary conditions, see e.g. [7]

(6) $$B^{r+1}(x) = \ldots = B^s(x) = 0, \text{ on } g, \text{ where}$$

(7) $$(Au,v) - \langle u,v \rangle = \int_g (B_1(v) B^1(u) + \ldots + B_s(v)\, B^s(u))\, dt,$$

is a generalized Green formula obtained by partial integration, $A^{1/2}$ is an isometry from X^o to $L_2(G)$

$$(A^{1/2} x,\ A^{1/2} y) = \langle x,y \rangle,\ (Ax,y) = \langle x,y \rangle, \text{ when } x \in X^o,\ y \in X.$$

The Green function $G = G(t,s)$, of A corresponding to (2) satisfies

(8) $$\langle G(\circ, t),x\rangle = (t) = f_t(x), \quad x \in X \; ||f||_{X^*} = G(t,t)^{1/2}$$

$$\langle G(\circ,s), G(\circ, t)\rangle = G(s,t)$$

and the complete set of boundary conditions, (2), (6), is a reproducing kernel in X.

Now there exists a stochastic process, (generalized) "Gaussian random field", Y, defined over G and associated to the space X, through its covariance

(9) $$E(y(t) \,.\, y^*(s)) = G(t,s), \quad t,s \in G, \; E\, y(t) \equiv 0$$

Such fields are Markovian, a.e. $(m - \frac{k}{2})$ times continuously differentiable and under some additional conditions, e.g. $m > k$, see [5] - which assures that $C^{1/2}$ is nuclear - satisfy a stochastic (pseudo) differential equation

$$A^{1/2}\, Y = M_Y, \text{ i.e. } (Y, A^{1/2}\, \phi) = W_Y(\phi), \text{ for } \phi \in C_o^\infty(G), \text{ a.e.,}$$

where W_Y is a "white noise", (depending only on Y), i.e.

$$W_Y(u) = \sum_j Z_j(u,u_j), \; Y(\phi) = \Sigma Z_j(\phi, A^{-1/2}\, u_j)$$

W_Y is a unitary map $L^2(G) \to Y \_ L_2(\Omega,F,P)$, where $\{Z_j\}$ is an orthonormal basis in Y, (uniquely) corresponding to an (arbitrary) ortonormal basis $\{U_j\}$ in $L_2(G)$, satisfying the complete set of boundary conditions (2), (6), (take e.g. the eigen functions of C). For the validity of the representation as well as for (11), (13), the above mentioned additional assumptions (used for (9)) are not needed. The values of f(y), where f is a linear continuous functional over X, (identified with an element of X by $f(x) = \langle f,x\rangle$) are defined as realizations of the random variable

(11) $$Y(f) = \sum_j Z_j\; (f,A^{-1/2}u_j), \text{ then}$$

$$E(f^2(y)) = \sum_j |(f,A^{-1/2}u_j)|^2 = \sum_j |(A^{-1/2}f,u_j)|^2 = ||A^{-1/2}\; f||_o^2 = ||f||^2.$$

Suppose now that a strongly compact set, M (of allowed measurements, f_m) of the dual space, X^*, is given together with a finite, positive measure, s(dm) on M, such that the map, $C = B{*}B$

(12) $$\langle Bx, Bx\rangle = \int_M \langle f\;, x\rangle^2 \; s(dm),$$

is nuclear. A special case, of main interest for us, is when $M \equiv G$ according to (4), (5), s is the Lebesgue measure on G, C will be nuclear by the assumption $2m > k$. The nuclearity of C is assumed in order that - by Fubini's theorem

$$(13) \qquad E\left(\int_M f_m^2(y)\, s(dm)\right) = \int_M \|f_m\|^2 s(dm) = \int_M \Sigma\langle f_m, E_i\rangle^2 \; s(dm) =$$

$$= \Sigma\langle CE_i, E_i\rangle = \text{Spur } C < \infty$$

$$CE_i = e_i E_i \;, \; AE_i = e_i^{-1} E_i, \; \langle E_i, E_i\rangle = 1, \quad i = 1,2,\ldots$$

<u>The optimal approximation (filtering) problems</u>

Suppose now that f, $f_1, \ldots, f_N$ are linear, continuous functionals on X and we have to estimate the value of $f(x) = \langle f,x\rangle = c$ in terms of the values of $c^N = (c_1, \ldots, c_N)$, $f^N = (f_1, \ldots, f_N)$, $f_i(x) = c_i$, $i = 1,\ldots,N$. We are interested, thus to find the optimal operator, $\overline{F}_d$ and

$$(14) \; E_d(f,f^N) = \inf_F \sup_{\|x\|\le 1} \{|f(x) - F(c^N,f^N,f)| \,|\, f_i(x) = c_i, \; i = 1,\ldots,N, x\in X\}$$

In a corresponding stochastic problem, using the Gaussian property (the best mean square estimation, the conditional mean, is a <u>linear</u> function of the measured values), we have to find

$$(15) \qquad \begin{aligned} &E_y\{f(y)\,|\,f_i(y) = c_i, \; i = 1,\ldots,N\} = \sum_i \overline{a}_i f_i(y) = \overline{F}_s(c^N,f^N,f) \\ &E_s(f,f^N) = \inf_{a^N} \; E_y\{(f(y) - \sum_i a_i \; f_i(y))^2\}. \end{aligned}$$

For the approximation of Bx in the deterministic case we shall study the following optimal sensor location problem

$$(16) \qquad \inf_{f^N\in M^N} \; \inf_F \; \sup_{\|x\|\le 1} \{\|x - F(c^N,f^N)\|_{L^2(M,s)} \,|\, f_i(x) = c_i, \; i = 1,\ldots,N, \; x\in X\} = e_d(N).$$

We remark that (16) turns out to be equivalent, see (23), (32) below, to a problem of optimal structural design, e.g. support points for an elastic plate.

The corresponding stochastic optimal sensor location problem is

$$(17) \qquad \inf_{f^N\in M^N} \; E\;\left(\int_M |f_m(y) - E(f_m(y)| \; f_i(x) = c_i, \quad i=1,\ldots, N)|^2 \; s(dm)\right)$$

where the outer expectation is extended over all possible realizations of c^N.

Remark. One could extend the above class of problems to inclued approximation of vector functions x, given by

$$\mathring{x}(t) = A(t)x(t) + u(t),\ x(t)\in R^n,\ B_i(x(a), x(b)) = 0,\quad i = 1,\dots,r$$

(18) $$||x||^2 := x^*(t_o)A_o x(t_o) + \int_a^b (x^*(s)A_1(s)x(s) + u^*(s)R(s)u(s))ds \le 1,$$

$$a \le t_o \le b,\ f_i(x) = H(t_i)y(t_i),\ \langle Cx,x\rangle = \int_a^b x^*(s)C_2(s)x(s)ds,\ a \le t_i \le b$$

and their stochastic counterparts, for $A_1 \equiv 0$, driven by white noise

(19) $$\mathring{y}(t) = A(t)y(t) + \omega(t),\ E_y(t) \equiv 0,\ B_i(y(a), y(b)) = 0,\quad i = 1,\dots,r$$

$$E\ (\omega^*(t)\ \omega(s)) = R(s)\ \delta(t-s),\ E(x(t_o)\ x(t_o)) = A_o$$

In fact, the "duality theorem", first formulated in [2] as the equivalence of minimum norm control and Gaussian filtering problems for dual pairs of linear systems, is generalized in our Theorem 1 to distributed parameter systems by also exhibiting the intuitive background (the relations (8) - (9)) of why such an equivalence holds, see also [6].

Exact formulas, characterizations and errors bounds for the optimal operators

Theorem 1. In both cases (14), (15) the optimal interpolation filtering operators and their errors are the same

(20) $$E_d^2\ (f,f^N) = E_s(f,f^N) = \langle f,f\rangle - \sum_{i,j=1}^{N} \langle f,f_j\rangle(\langle f_k,f_l\rangle)^{-1}_{ij} \langle f,f\rangle$$

(21) $$\overline{F}_s(f,\overline{f}^N,c^N) = \overline{F}_d(f,f^N,c^N) = \sum_{i,j=1}^{N} (\langle f_k,f_l\rangle)^{-1}_{ij}\langle f,f_i\rangle\ c_j = f(S(f^N,c^N)),$$

Notice that in the case of points measurements, (4), $\langle f_i,\ f_j\rangle = G(t_i,\ t_j)$ this everything can be expressed in terms of the Green function.

<u>Proof</u>. In the deterministic part we use the following geometric (and all of them alone already characteristic!) properties of ellipsoids: their planar sections (here of codimension N) are centralsymetrical and the centre of a planar section is the minimum norm element - (in the norm where the ellipsoid is the unit ball) - in that plane, the centres of parallel sections (as c^N varies) lie on a linear manifold, which means that these best, minimum norm or spline approximations are linear in c^N. For computing them we have to solve

$$\inf \{<x,x> \mid f_i(x) = c_i,\ i = 1,\dots,N,\ x \in X\}$$

The necessary (and here obviously sufficient) conditions in the form of Lagrange, see e.g. [9], with multiplicators, $a_1,\dots,a_N$, gives

$$\bar{x} = a_1 f_1 + \dots + a_N f_N,\ <f_i,\ x> = c_i,\ i = 1,\dots,\ N, \text{ thus}$$

(22)
$$\bar{x} = \sum_{i,j} (<f_k,\ f_l>)^{-1}_{ij}\ f_i\ c_j = S\ (f^N,\ c^N)$$

here the matrix is not singular by the linear independence of $\{f^N\}$. Notice that (22) is - in the problems (1) - (4) in fact a boundary value problem: x satisfies (2) and (6) and

$$Ax(s) = a_1\delta(t_1-s) + \dots + a_N\delta(t_N-s),\ x(t_i) = 0,\ i = 1,\dots,N.$$

In the case of the problem (18), (22) contains the Hamiltonian system consisting of (18) and

$$\dot{z}(t) = -A^*(t)z(t) + 2A_1(t)\ x(t) + \sum_{i=1}^{T} a_i H^*(t_i)\delta(t-t_i)$$

completed by the "transversality conditions", (like (6)), at a and b, see e.g. [9]. To prove (22) we notice that the largest error is realized for $c^N = 0$, thus we have to compute

(23) $\quad e = \sup \{<f,x> \mid <f_i,\ x> = 0,\ i = 1,\dots,N,\ <x,x> \le 1\}$

The Euler-Lagrange equation here yields, by $<\bar{x},\bar{x}> = 1$,

$$e\bar{x} = f + \Sigma a_j f_j,\ <f_i,\ \bar{x}> = 0,\ i = 1,\dots,N, \text{ thus}$$

$$e = <f,\bar{x}> = \frac{1}{e} <f,f> - \frac{1}{e} \sum_{i,j}\ <f,f>)^{-1}_{j,i} <f_j,f>\ <f_i,f>,$$

by simple computation. The stochastic parts of the relations (20), (21) are well known, (proved elementary by minimizing a quadratic form).

The above mentioned geometric properties show that also for the approximation of Bx, the spline, $S(f^N, c^N)$, is the optimal approximation operator (in fact, for any Banach space norm over BX, just by the centralsymmetriy of sections).

In the stochastic case the corresponding relation

$$E_y(B\, y \mid f_i(y) = c_i,\quad i=1,\dots,N) = B\, S(f^N, c^N)$$

is a consequence of the onedimensional case, (21), by expanding y into eigenfunction series of the operator B, (C)). - Going over to the optimal sensor location problems (16), (17), first notice that, for fixed f^N, the global error is easily computed in the problem (17), (by simple integration)

$$(24)\qquad E_s(Bx,\ f^N) = \mathrm{Spur}\ C - \sum_{i,j=1}^{N} (\langle f_k, f_l\rangle)^{-1}_{ij} \langle f_i,\ Cf_j\rangle$$

In the problem (16), for fixed f^N, for the largest error e, realized, by $\bar{x}$, for $c^N = 0$

$$(25)\qquad e^2\bar{x} = C\,\bar{x} - a_1 f_1 - \dots - a_N f_N, \langle f_j,\ x\rangle = 0,\ j=1,\dots,N,\ \langle\bar{x},\bar{x}\rangle = 1$$

the Euler-Lagrange equation, see e.g. [9], (for a "structural optimization problem", like (23)), yields

$$(26)\qquad E_d(Bx, f^N) = \sup\{a \mid \det\langle f_j, (I-\frac{1}{a^2}\ C)^{-1}\ f_k\rangle_{j,k=1,\dots,N} = 0\},$$

at least when $\bar{a}^2$ is not an eigenvalue of C, e.g. $\{f^N\}$ do not lie on the zero set of an eigenfunction. Let the eigenvalues of $C = B^2$ ordered decreasingly as follows

$$(27)\qquad e_1 \ge e_2 \ge \dots \ge e_q > e_{q+1} = \dots = e_N = \dots = e_{N+m} > e_{N+m+1} \ge \dots,$$

<u>Theorem 2.</u> For arbitrary $f_1, \dots, f_N$

$$(28)\qquad E_s(Bx, f^N) \ge \sum_{j\ge 1} e_{N+j}(B^2),\ \text{here equality holds if and only if } L(f^N) = L(E^N)$$

i.e. the subspace of the first N eigenfunctions is the unique (modulo the ordering in (27)) optimal set of N linear sensors.

Proof. By the formula (24) we have to prove that

$$\mathrm{Spur}\ (G^{-1}\ G_c) \le e_1(C) + \dots + e_N(C)\ ,$$

where G, G_c are the Gram matrices of f^N. Let us denote the orthogonal projection of X to the subspace $L(f^N)$ by P_N, and T the transformation which takes f^N into an orthonormed set of vektors, i.e. $T\ G\ T^* = I$, then the matrix of the map, $C_N = P_N C P_N$, defined over $L(f^N)$ is easily seen to be $T\ G_c\ T^*$, therefore

$$C_N = (T^*)^{-1}\ G^{-1}\ G_c\ T^*, \text{ thus Spur } C_N = \mathrm{Spur}\ (G^{-1}\ G_c).$$

Now Spur $C_N = \sum_i e_i \langle p_i, p_i\rangle$, where $p_i = P_N E_i$, i=1,..., and

$N = \mathrm{Spur}\ P_N = \sum_i \langle p_i, p_i\rangle$, and $\langle p_i, p_i\rangle \le 1$ yields

$\sum_i e_i \langle p_i, p_i\rangle \le e_1 + \dots + e_N$, with equality iff $\langle p_i, p_i\rangle = 1$,

i=1,...,N.

Theorem 3. The necessary and sufficient condition for the attainment of the (exact) lower bound

(29) $E_d(Bx, f^N) \ge \sqrt{e_{N+1}(B^2)}$, is that index $M_N\ (e_{N+1}) = q$,

where $M_N(a) = \langle f_j, (I - \frac{1}{a} B^2)^{-1}\ f_k\rangle_{j,k=1,\dots,N}$.

A necessary condition for the attainment of the equality in (29) is that $L(f^N)$ must be orthogonal to some m-dimensional subspace of $L(E_{q+1}, \dots, E_{N+m})$, and

(30) $$\mathrm{rank}\ (\langle f_j, E_k\rangle)_{j=1,N,k=1,\dots,q} = q$$

and under the latter conditions a sufficient condition for the equality in (29) is that

(31) $$\sum_{k=1}^{N} e_k\ d^2(E_k, (f^N)) \le e_{N+m+1} - e_N$$

Proof. The inequality (29) was first proved in a simple case, in [1] (contrary to what was expected in [1]) it turns out that equality holds in (29) not only for the choice $L(f^N) = L(E^N)$, see the example below. The condition is proved simply by the observation that

$D = (I - s\,B^2)$ is nonnegative definit over the zero space, $N(f^N)$ if and only if its index (number of negative eigenvalues) over the whole space X is equal to its index over the subspace $L(D^{-1}\,f^N)$. Which is D orthogonal to $N(f^N)$, (we write $\langle f,x\rangle = \langle D^{-1}\,f,\,Dx\rangle$).

For this, the criterion, (see [7] for a complicated proof), is well known: in the sequence $W_i(a) = \det M_i(a)$, $i=1,\ldots,N$, there are exactly q changes of sign for all $a = e_{N+1}-\varepsilon$, ε small enough (30) is proved by noting that for linear combinations $p_1E_1 + \ldots + p_qE_q$ the Rayleigh quotient $||Bx||^2/||x||^2$, is larger than e_{N+1}, thus they should not be in $N(f^N)$. The orthogonality condition and (31) have been proved in [8].

The essential nonuniquess of the optimal sets of nodes for M = G, (4), can be checked in the example, where G is a rectangular "plate", with siedes of length, m, l, simply supported on the boundary, here $A = \Delta^2$ the biharmonic operator, $B_1(x) = x$, $B^2(x) = \Delta x$, if $m \ll \ell$ the points (t^N) near the centres of the N equidistant segments parallel to the side m, are all optimal. The matrices M_N can be computed explicitly, e.g. by the formula, (where $||E_n||_\circ = 1$)

$$(32)\qquad \langle f_j,(I - \tfrac{1}{a}\,B^2)^{-1}\,f_k\rangle = \sum_{n=1}^{\infty} E_n(t_j)\cdot E_n(t_k)\,(e_n^{-1} - a)^{-1}$$

which can be evaluated approximately by the matries of the corresponding one dimensional problem, where the optimality of the unique system of N zeros of the (N+1)-st eigenfunctions is checked directly, by completing to the square, like in Picone's inequality

$$(33)\qquad \int_a^b (a_1(t)(x'(t))^2 + a_o(t)x^2(t) - e_{N+1}^{-1}x^2(t))dt =$$

$$= \int_a^b p_1(x' - E'_{N+1}\frac{x}{E_{N+1}})^2\,dt,$$

where $-(a_1(t)\,E'_{N+1}(t))' + a_o(t)E_{N+1}(t) = e_{N+1}^{-1}\,E_{N+1}(t)$,

$E_{N+1}(a) = E_{N+1}(b) = 0$,

(a simplest case, m = k = 1 of our problems). In the general, m-th order onedimensional problem the existence and optimality of the (unique) system of N-nodes of the (N+1)-st eigenfunction should be proved by the extensions of the methods of Morse theory.

Remark. The most important feature of the Kalman filter is its recursivity, our splines (and their nodes for $N\to\infty$, see [10]) could also be computed recursively in N, yet, as noted above no sequential choice of the measurements f_i, $i=1,\dots,N$ can improve the global error (in fact even in the local sense they cannot be more accurate). On the other hand, e.g. in the problem of uniform approximation (in C(a,b)) of functions with convex r-th derivatives (r arbitrary) on [a,b], for fixed values of $x^{(j)}(a)$, $x^{(j)}(b)$, $j=0,\dots,r+1$, there exists sequential algorithms for the choice of t^N which are $c_r \cdot N$-times more accurate - for all N - than the optimal passive ones, see [10], and the same holds for the corresponding errors (i.e. variances) if the (r+1)-st derivatives are assumed to be random, piecewise constant, monoton functions, whose jump points, in [a,b], and (jump) values, in $[x^{(r+1)}(a), x^{r+1}(b)]$ are independent Poisson processes.

References

[1] Kolmogorov, A.N., Über die beste Annäherung von Funktionen einer gegebenen Funktionenklasse, (1936), Annals of Math. 37, 107-110.

[2] Kalman, R.E., Bucy, R.C., New results in linear filtering and prediction theory, Journ. Basic Eng. (1961), 83, 95-108.

[3] Ciesielski, Z., Probabilistic and analytic formulas for the periodic spline interpolating with multiple nodes, (1979), Banach Center Publications, vol. 5, 35-47.

[4] Surgailis, D., On trajectories of Gaussian random fields. Banach Center Publications, (1979), vol. 5. 231-247.

[5] Yosida, K., Functional Analysis, (1966), Springer Verlag, New York.

[6] Kuržanskii, A.B., Control and Observation under Uncertainty, (in Russian), 1977. Nauka, Moscow.

[7] Gould, S.H., Variational Methods for Eigenvalue Problems. (1966), Oxford University Press.

[8] Karlowicz, L.A., Remarks on Variational Characterizations of Math. Anal. and Appl., (1976), 53, N° 1, 99-111.

[9] Gamkrelidze, R.V., Necessary conditions of extrema, Trudi Inst. Steklova, (1971), (in Russian), vol. CXII., 152-180.

[10] Sonnevend, Gy., Uniform,sequential,N-step approximation of functions with convex r-th derivatives, Analysis Mathematica (to appear).

Dept. of Numerical Math.
Eötvös University
1445, Budapest, 8. Pf. 323
Hungary

ON FIRST ORDER SINGULAR BELLMAN EQUATION

H. Pragarauskas

Institute of Mathematics and Cybernetics
Academy of Sciences of the Lithuanian SSR
Vilnius

In the paper the controlled stochastic processes, consisting of drift and jump terms are considered. The cost function is characterized as the unique solution of the Bellman equation, which is in this case a non-linear integro-differential first order equation.

1. The main results

Let R^d be a d-dimensional Euclidean space, $T\in(0,\infty)$, $H_T = [0,T]\times R^d$, A a separable metric space, Π a measure on R^d with the differential $dz/|z|^{d+1}$ and $\mathcal{L}$ a class of functions $u : R^d \to \{y\in R^d, |y| \le 1\}$ such that $||u|| \equiv \int|u(z)|\Pi(dz) < \infty$.

Let for all $(t,x)\in H_T$, $\alpha\in A$ be defined: $b(\alpha,t,x)\in R^d$, $c(\alpha,t,x,\cdot)$ an element of $\mathcal{L}$ and real $r(\alpha,t,x) \ge 0$, $f(\alpha,t,x)$, $g(x)$.

We shall use the following assumptions.

I. b,r,f,g are Borel measurable in t, continuous in α and continuous in x uniformly with respect to α for any t, c is Borel measurable in (α,t,x,z) and for any $\alpha\in A$, $(t,x)\in H_T$ $||c(\alpha,t,x,\cdot) - c(\beta,t,x,\cdot)|| \to 0$ as $\beta \to \alpha$.

II. b is bounded and Lipschitz continuous in x uniformly with respect to (α,t), $||c||$ is bounded and c is Lipschitz continuous in x (in sense of norm $||\cdot||$) uniformly with respect to (α,t).

III. r,f,g are bounded and Lipschitz continuous in x uniformly with respect to (α,t).

IV. For any $\alpha\in A$ there exists a measure $\overline{\pi}(\alpha,dy)$, $\int|y|\overline{\pi}(\alpha,dy) < \infty$ on Borel subsets of $\{y\in R^d, |y| \le 1\}$ such that for all $(t,x)\in H_T$ the measure $\pi(\alpha,t,x,dy) \equiv \Pi(z : c(\alpha,t,x,z)\in dy\setminus\{0\})$ is absolutely continuous with respect to the measure $\overline{\pi}(\alpha,dy)$ and

$$\operatorname*{esssup}_{H_T} \int[p(\alpha,t,x-y,y) - p(\alpha,t,x,y)]\overline{\pi}(\alpha,dy) < \infty,$$

where $\quad p(\alpha,t,x,y) = \pi(\alpha,t,x,dy)/\overline{\pi}(\alpha,dy)$.

Let $(\Omega,\mathcal{F},P)$ be a complete probability space with a family $(\mathcal{F}_t)$ of complete non-decreasing σ-algebras $\mathcal{F}_t \subset \mathcal{F}$, $(z_t, \mathcal{F}_t)$ a d-dimensional Cauchy process with a Levy measure Π and p a random Poisson measure on $[0,\infty)\times R^d$ constructed from the jumps of z_t.

Let $\mathcal{A}$ be a class consisting of all processes $\alpha_t : \Omega \times [0,T] \to A$ progressively measurable with respect to $(\mathcal{F}_t)$.

If assumptions I,II are satisfied, then for any $(s,x)\in H_T$, $\alpha\in\mathcal{A}$ there exists the unique solution $x_t^{\alpha,s,x}$ of the Ito equation

$$x_t = x + \int_0^t b(\alpha_r,\ s + r,\ x_r)\,dr + \int_0^t\!\!\int c(\alpha_r,\ s + r,\ x_r,z)p(drdz).$$

Let the cost function v be defined by the formula

$$v(s,x) = \sup_{\alpha\in\mathcal{A}} E_{s,x}^{\alpha}\Big[\int_0^{T-s} e^{-\varphi_t} f(\alpha_t,s+t,x_t)dt + e^{-\varphi_{T-s}}\, g(x_{T-s})\Big],$$

where

$$\varphi_t^{\alpha,s,x} = \int_0^t r(\alpha_u,s + u,\ x_u^{\alpha,s,x})du.$$

Here and below for convenience of notation the indexes α,s,x are written at the symbol of an expectation.

Let $\Lambda(\Lambda_{loc})$ be a class of bounded (locally) and Lipschitz continuous (locally) functions on H_T. If $u\in\Lambda_{loc}$ then locally bounded partial derivatives u_t, u_{x_i} exist in Sobolev sense.

Define the operators L^α, F^α, F for $\alpha\in A$ acting on Λ_{loc} by the formulae:

$$L^\alpha u(t,x) = u_t(t,x) + \sum_{i=1}^{d} b_i(\alpha,t,x)u_{x_i}(t,x) + \int[u(t,x + y) -$$

$$- u(t,x)]\ \pi(\alpha,t,x,dy) - r(\alpha,t,x)u(t,x),$$

$$F^\alpha u(t,x) = L^\alpha u(t,x) + f(\alpha,t,x), \quad Fu(t,x) = \sup_{\alpha\in A} F^\alpha u(t,x).$$

<u>Theorem 1.1.</u> Let the assumptions I-III hold. Then $v\in\Lambda$ and $Fv = 0$ a.e. H_T, $v(T,\cdot) = g$.

If Q is a subset of R^k then by $L^p(Q)$ we shall denote a class of function u on Q such that $||u||_{p,Q}^p \equiv \int_Q |u(x)|^p dx < \infty$ and by $L_{loc}^p(Q)$ a class of functions u on Q such that $u \in L^p(Q_1)$ for every compact subset $Q_1 \subset Q$.

Let Λ^+ be a class of functions u on H_T such that the following conditions are satisfied:

(i) $u = u' + u''$, $u', u'' \in \Lambda_{loc}$,

(ii) there exist constants $N \geq 0, n \geq 0$ such that for all $(t,x) \in H_T$

$$(1.1) \qquad |\eta(t,x)| \leq N(1 + |x|)^n, \quad \eta = u', u'',$$

(iii) there exists a constant $\gamma \in (1,2)$ such that

$$(1.2) \qquad \int_{|y| \leq 1} \frac{|\Delta_y^2 u'(t,x)|}{|y|^{d+\gamma}} dy \in L^1{}_{loc}(H_T), \quad \int_{|y| \leq 1} \frac{\Delta_y^2 u'(t,x)}{|y|^{d+\gamma}} dy \geq 0 \text{ a.e.} H_T,$$

$$\int_{|y| \leq 1} \frac{|\Delta_y^2 u''(t,x)|^2}{|y|^{d+2\gamma}} dy \in L^p{}_{loc} (H_T) \text{ for some } p > \frac{d + \gamma}{\gamma - 1},$$

where $\Delta_y^2 u(t,x) = u(t,x + y) + u(t,x-y) - 2 u(t,x)$.

<u>Theorem 1.2.</u> Let the assumptions I, II, IV hold, f satisfies (1.1) uniformly with respect to $\alpha, u_1, u_2 \in \Lambda^+, Fu_1 = Fu_2 = 0$ a.e. H_T, $u_1(T,\cdot) = u_2(T,\cdot)$ on R^d. Then $u_1 = u_2$ on H_T.

<u>2. The existence theorem</u>

In this section we shall prove Theorem 1.1.

<u>Lemma 2.1.</u> Let the assumptions I-III hold. Then $v \in \Lambda$.

<u>Proof.</u> Using the assumptions II, III it is not difficult to prove that v is bounded and Lipschitz continuous in x uniformly with respect to t. Using an approach similar to the proof of Theorem III.3.6 [2] one can derive, that for any $(s,x) \in H_T$ and $t \in [0,T-s]$

$$(2.1) \qquad v(s,x) = \sup_{\alpha \in \mathcal{A}} E_{s,x}^{\alpha}\left[\int_0^t e^{-\varphi_r} f(\alpha_r, s+r, x_r) dr + e^{-\varphi_t} v(s+t, x_t)\right].$$

Now using the assumption III, Lipschitz continuity in x of function v and (2.1) we derive, that for any $(s,x)\in H_T$, $t\in[0,T-s]$

$$|v(s,x) - v(s+t,x)| \le \sup_{\alpha\in\mathcal{A}} E^{\alpha}_{s,x}\Big|\int_0^t e^{-\varphi_r} f(\alpha_r, s+r, x_r)dr +$$

$$+ e^{-\varphi_t} v(s+t,x_t) - v(s+t,x)\Big| \le \text{const. } t.$$

This completes the proof.

Denote by $C_0^{1,1}$ a class of continuously differentiable functions on H_T with compact support.

The proof of following two lemmas is analoguous to the proof of Lemmas 1.2 [3], 1.3 [3] respectively.

Lemma 2.2. Let the assumptions I-III hold, $u\in C_0^{1,1}$. Then:

(i) $F^{\alpha}u(t,x)$ is continuous in α, continuous in x uniformly with respect to α and Borel measurable in t,

(ii) $Fu(t,x)$ is continuous in x and Borel measurable in t.

Lemma 2.3. Let $\Lambda(x)(\Lambda^{\beta}(x),\ \beta\in A)$ be the set of Lebesgue points of function $Fu(\cdot,x)$ $(F^{\beta}u(\cdot,x))$ for all $u\in C_0^{1,1}$. Then for any $x\in R^d$ the Lebesgue measure of the set $(0,T)\setminus\Lambda(x)$ $((0,T)\setminus\Lambda^{\beta}(x))$ is zero.

If $u\in\Lambda_{loc}$ then by Rademacher theorem for almost all $(s,x)\in H_T$

$$(2.2)\qquad u(s+t,x+y) = u(s,x) + u_s(s,x)t + \sum_{i=1}^{d} u_{x_i}(s,x)y_i + $$
$$+ o(|t| + |y|)$$

as $|t| + |y| \to 0$ $\qquad \left(\frac{o(r)}{r} \longrightarrow 0 \quad \text{as } r\to 0\right).$

Lemma 2.4. Let the assumptions I-III hold, $u\in\Lambda$, B is an arbitrary countable subset dense in A. Then $\sup_{\alpha\in A} F^{\alpha}u(s,x) = \sup_{\alpha\in B} F^{\alpha}u(s,x)$ for every point (s,x) for which (2.2) holds.

Proof. Fix a point (s,x) for which (2.2) holds. Let $\{u^n\}$ be a sequence of functions $u^n\in C_0^{1,1}$ such that for all $s+t\in[0,T]$, $y\in R^d$

$$|u(s+t,x+y) - u^n(s+t,x+y)| \le \frac{1}{n}(|t| + |y|).$$

Such sequence we can construct if we set

$$u^n(s+t,x+y) = u(s,x) + u_s(s,x)t + \sum_{i=1}^{d} u_{x_i}(s,x)y_i$$

for sufficiently small (t,y). Then using the continuity of u we can extend u^n to smooth function with compact support which is sufficiently close to u on sufficiently large cylinder. In this case $u^n(s,x) = u(s,x)$, $u^n_{x_i}(s,x) = u_{x_i}(s,x)$, $u^n_s(s,x) = u_s(s,x)$ and $F^\alpha u^n(s,x) \to F^\alpha u(s,x)$ as $n \to \infty$ uniformly with respect to α. Since $F^\alpha u^n(s,x)$ is continuous in α it follows that $F^\alpha u(s,x)$ is continuous in α. This completes the proof.

Proof of Theorem 1.1. By Lemma 2.1 $v \in \Lambda$. Fix $(s,x) \in H_T$, $s \in \Lambda(x)$ such that (2.2) holds for $u \equiv v$. Let $\{v^n\}$ be a sequence of functions $v^n \in C^{1,1}_o$ (constructed as in Lemma 2.4) such that for all $t \in [0,T-s]$, $y \in R^d$

$$|v(s+t,x+y) - v^n(s+t,x+y)| \le \frac{1}{n}(|t| + |y|) \tag{2.3}$$

and $v^n(s,x) = v(s,x)$, $v^n_{x_i}(s,x) = v_{x_i}(s,x)$, $v^n_s(s,x) = v_s(s,x)$.

From (2.1), (2.3) and well known estimates for moments of $x_t^{\alpha,s,x}$ it follows that

$$\begin{aligned} v^n(s,x) \le \sup_{\alpha \in \mathcal{A}} E^\alpha_{s,x}\Big[\int_0^t e^{-\varphi_r} f(\alpha_r, s+r, x_r)dr + \\ + e^{-\varphi_t} v(s+t,x_t)\Big] + \frac{t}{n} N(1+|x|), \end{aligned} \tag{2.4}$$

where a constant N is independent of t,n. Applying Ito formula to $e^{-\varphi_t} v^n(s+t,x_t)$ and using (2.4) we derive

$$0 \le \sup_{\alpha \in \mathcal{A}} E^\alpha_{s,x} \int_0^t e^{-\varphi_r} Fv^n(s+r,x_r)dr + \frac{t}{n} N(1+|x|) \le$$

$$\le \int_0^t Fv^n(s+r,x)dr + \sup_{\alpha \in \mathcal{A}} E^\alpha_{s,x} \int_0^t |Fv^n(s+r,x_r) - Fv^n(s+r,x)|dr +$$

$$+ \sup_{\alpha \in \mathcal{A}} E^\alpha_{s,x} \int_0^t |Fv^n(s+r,x_r)| \int_0^r r(\alpha_u, s+u, x_u)dudr + \frac{t}{n} N(1+|x|).$$

Dividing this inequality on t and using the continuity of $Fv^n(s+r,y)$ in y we obtain

$$0 \le \lim_{t \downarrow 0} \frac{1}{t} \int_0^t F\, v^n(s+r,x)\, dr + \frac{N}{n}(1+|x|). \tag{2.5}$$

Since $s \in \Lambda(x)$, $F^{\alpha}v^{n}(s,x) \to F^{\alpha}v(s,x)$ as $n \to \infty$ uniformly with respect to α (see the proof of Lemma 2.4) and in particular $Fv^{n}(s,x) \to Fv(s,x)$ as $n \to \infty$ from (2.5) it follows that $Fv(s,x) \geq 0$. This inequality and Lemma 2.3 imply $Fv \geq 0$ a.e. H_T.

Now we shall prove the converse inequality. Fix an arbitrary $\beta \in A$. Let a point $(s,x) \in H_T$ be such that (2.2) holds for $u \equiv v$ and $s \in \Lambda^{\beta}(x)$.

From (2.1), (2.3) and well known estimates for moments $x_t^{\beta,s,x}$ we obtain

$$v^{n}(s,x) \geq E_{s,x}^{\beta}\Big[\int_0^t e^{-\varphi_r} f(\beta, s + r, x_r)dr + e^{-\varphi_t} v^{n}(s + t, x_t) +$$

$$+ N\frac{t}{n}(1 + |x|),$$

where a constant N is independent of t,n. Further, using an approach similar to given above we conclude $F^{\beta}v \leq 0$ a.e. H_T. Taking here the upper bound by the countable subset dense in A and using Lemma 2.4 we derive $Fv \leq 0$ a.e. H_T. This completes the proof.

3. The unicity theorem

In this section we shall prove Theorem 1.2. This proof is based on the probabilistic representation of a solution of the Bellman equation. In the sequel we need the following definitions and results.

If Q is a subset of R^k then by $\bar{Q}$ we denote its closure and by ∂Q its boundary. $C_0^{\infty}(Q)$ is a class of infinitely differentiable functions on Q with compact support.

Let $\zeta \in C_0^{\infty}(R^{d+1})$ be a non-negative function, such that $\zeta = 0$ if $|x| \geq 1$ and $||\zeta||_{1,R^{d+1}} = 1$. If function u is locally integrable on R^{d+1}, then by $u^{(n)}$ we denote a convolution of u with the kernel $n^{d+1}\zeta(nx)$, $n = 1,2,\ldots$

Denote by $\mathcal{K}$ a set of all collections $\kappa = (\Omega, \mathcal{F}, \mathcal{F}_t, P, z_t, p(dtdz))$, where $(\Omega, \mathcal{F}, P)$ is a complete probability space with the family $(\mathcal{F}_t$, $t \geq 0)$ of complete non-decreasing σ-algebras $\mathcal{F}_t \subset \mathcal{F}$, $(z_t, \mathcal{F}_t)$ is a d-dimensional Cauchy process with Levy measure Π and p is a Poisson random measure, constructed from the jumps of z_t. The elements of a collection κ we shall denote by Ω^{κ}, $\mathcal{F}^{\kappa}$, ..., $p^{\kappa}(dtdz)$.

Fix an arbitrary $\kappa \in \mathcal{K}$, $\beta \in A$ and denote

$$\lambda = \operatorname*{esssup}_{H_T} [-\sum_{i=1}^{d} b_{ix_i}(\beta,s,x) + \int [p(\beta,s,x-y,y)-p(\beta,s,x,y)]\overline{\pi}(\beta,dy)],$$

$$\Pi_t^{\beta} g(s,x) = E_{s,x}^{\kappa,\beta}\, g(x_{t-s}),\quad R_t^{\beta} h(s,x) = E_{s,x}^{\kappa,\beta} \int_0^{t-s} h(s+r,\, x_r)dr,$$

where $t \in [s,T]$, $g \in L^p(R^d)$, $h \in L^p(H_t)$ for some $p \geq 1$ and $x_t^{\kappa,\beta,s,x}$ is the unique solution of the equation

$$x_t = x + \int_0^t b(\beta,s+r,\, x_r)dr + \int_0^t \int c(\beta,s+r,\, x_r,\, z)\, p^{\kappa}(drdz).$$

Lemma 3.1. Let the coefficients $b(\beta,s,x)$, $c(\beta,s,x,z)$ satisfy the assumptions I,II. Then for every $g \in L^p(R^d)$, $h \in L^p(H_t)$, $p \geq 1$

(3.1) $$||\Pi_t^{\beta} g(s,\cdot)||_{p,R^d}^p \leq e^{\lambda(t-s)} ||g||_{p,R^d}^p ,$$

(3.2) $$||R_t^{\beta} h||_{p,H_t}^p \leq t^p e^{\lambda t} ||h||_{p,H_t}^p .$$

In particular the functions $\Pi_t^{\beta} g(s,x)$, $R_t^{\beta} h(s,x)$ are defined for almost all $x \in R^d$, $(s,x) \in H_t$ respectively.

Proof. First we shall note that applying Hölder inequality it is not difficult to derive (3.2) from (3.1). Hölder inequality also implies, that it is sufficient to prove (3.1) for $p = 1$, $g \geq 0$, $g \in C_0^{\infty}(R^d)$. In this case by theorem 1.1 for $u(s,x) \equiv \Pi_t^{\beta} g(s,x)$ we have a.e. H_t

$$u_s(s,x) + \sum_{i=1}^{d} b_i(\beta,s,x)u_{x_i}(s,x) + \int [u(s,x+y) - u(s,x)]\pi(\beta,s,x,dy) = 0.$$

Let us multiplicate this equality on $\psi\eta$, where $\psi,\eta \geq 0$, $\psi \in C_0^{\infty}(R^d)$, $\eta \in C_0^{\infty}((0,t))$. Then integrating on H_t and using Fubini theorem we obtain

$$- \int_{H_t} u\, \{\psi\eta_s + \sum_{i=1}^{d} (b_i\psi)_{x_i}\eta - \eta\int[\psi(x-y)p(\beta,s,x-y,y) -$$

(3.3)

$$- \psi(x)p'(\beta,s,x,y)]\overline{\pi}(\beta,dy)\}\, dsdx = 0.$$

Since g has a compact support and b,c are bounded it follows that $\lim_{|x|\to\infty} u(x)|x|^n = 0$ for every $n > 0$. Now replacing ψ in (3.3) by the function $\psi(\frac{x}{R})$ such that $\psi(0) = 1$ and tending $R \to \infty$ we derive

$$-\int_{H_t} u\{\eta_s + \sum_{i=1}^{d} b_{ix_i}\eta - \eta\int[p(\beta,s,x-y,y)-p(\beta,s,x,y)]\bar{\pi}(\beta,dy)\, dsdx = 0.$$

Consequently, for every $\eta \geq 0$, $\eta \in C_0^\infty((0,t))$

$$\int_{H_t} u(\eta_s - \lambda\eta)dsdx \leq 0.$$

The rest of the proof is the same as the corresponding part of the proof of Lemma 2.1 [1].

Fix an arbitrary $\kappa \in \mathcal{K}$, $\gamma \in (1,2)$. Let $(\xi_t, \mathcal{F}_t^\kappa)$ be a homogeneous process independent of z_t^κ with the independent increments and the characteristic function

$$(3.4)\quad E^\kappa e^{i(\theta,\xi_t)} = \exp\{-\frac{1}{2}t\int_{|y|\leq 1}[e^{i(\theta,y)} + e^{-i(\theta,y)} - 2]\frac{dy}{|y|^{d+\gamma}}\}.$$

Let $b : [0,T]\times\Omega^\kappa \to R^d$, $c : [0,T]\times\Omega^\kappa \times R^d \to R^d$ be processes progressively measurable with respect to $(\mathcal{F}_t^\kappa)$ and such that for some constant K for all $(t,\omega) \in [0,T]\times\Omega^\kappa$

$$|b(t,\omega)| + ||c(t,\omega,\cdot)|| \leq K.$$

Fix $\delta > 0$ and denote

$$x_t = \int_0^t b(s)ds + \int_0^t\int c(s,z)p^\kappa(dsdz) + \delta\xi_t.$$

<u>Lemma 3.2.</u> For every $h \in L^p(H_T)$, $p > \frac{d+\gamma}{\gamma-1}$

$$E^\kappa \int_0^{T-s} h(s+t, x+x_t)dt \leq N\delta^{-1-\frac{d}{p}}||h||_{p,[s,T]\times R^d},$$

where a constant N depends only on K,T,γ,p,d.

<u>Proof.</u> It is sufficient to prove this lemma for $h \geq 0$, $h \in C_0^\infty(H_T)$. Denote

$$u^\delta(s,x) = \int_0^\infty e^{-t}\int_{R^d} h(s+t, x+\delta y)p(t,y)dydt,$$

where $p(t,y) = \frac{1}{(\sqrt{2\pi})^d} |y|^{-\frac{d}{2}+1} \int_0^\infty \rho^{\frac{d}{2}} \mathcal{J}_{\frac{d}{2}-1}(\rho|y|) e^{-t\rho^\gamma} d\rho,$

$\mathcal{J}_{\frac{d}{2}-1}$ is the Bessel function. u^δ is a resolvent of the process $\delta\eta_t$, where η_t is a stable process of an order γ. It is well known, that

$$u^\delta - u_t^\delta - c_\gamma \delta^\gamma L_\gamma u^\delta = h, \tag{3.5}$$

where

$$L_\gamma u(t,x) = \frac{1}{2} \int_{|y|\le 1} \Delta_y^2 u(t,x) \frac{dy}{|y|^{d+\gamma}} + \int_{|y|>1} [u(t,x+y)-u(t,x)] \frac{dy}{|y|^{d+\gamma}},$$

$$c_\gamma = 2^{\gamma-1} \pi^{-(\frac{d}{2}+1)} \sin \frac{\pi\gamma}{2} \cdot \Gamma(\frac{d+\gamma}{2}) \Gamma(\frac{\gamma+2}{2}).$$

Applying Hölder inequality it is not difficult to derive, that for some constants N_1, N_2 depending only on p,d,γ:

$$|u^\delta(s,x)| \le N_1 \delta^{-\frac{d}{p}} \{\int_0^\infty e^{-t} ||h(s+t,\cdot)||^p_{p,R^d} dt\}^{\frac{1}{p}}, \quad \text{if } p > \frac{d+\gamma}{\gamma}, \tag{3.6}$$

$$|u^\delta_{x_i}(s,x)| \le N_2 \delta^{-1-\frac{d}{p}} \{\int_0^\infty e^{-t} ||h(s+t,\cdot)||^p_{p,R^d} dt\}^{\frac{1}{p}}, \quad \text{if } p > \frac{d+\gamma}{\gamma-1}. \tag{3.7}$$

Since $h \ge 0$, $h \in C_0^\infty(H_T)$ it follows that u^δ is a non-negative infinitely differentiable function. Applying Ito formula to u^δ and using (3.5) we derive

$$E^K \int_0^{T-s} h(s+t, x+x_t)dt \le e^{T-s} E^K \int_0^{T-s} e^{-t} h(s+t, x+x_t)dt \le$$

$$\le e^{T-s}[u^\delta(sx) + \sup_{(t,x)} |\text{grad}_x u^\delta(t,x)| (2K + c_\gamma \delta^\gamma \int_{|y|>1} \frac{dy}{|y|^{d+\gamma}})] \cdot$$

The estimates (3.6), (3.7) complete the proof.

Using an approach similar to the proof of Theorem 2.1 [4] and applying Lemma 3.2 one can derive the following lemma.

<u>Lemma 3.3.</u> Let $(s,x)\in H_T$, $\delta > 0$, $\gamma\in (1,2)$, $b: [0,T]\times R^d \to R^d$, $C : [0,T]\times R^d\times R^d \to S_1$ are Borel measurable functions, such that for some constant K and all $(t,x)\in H_T$

$$|b(t,x)| + ||c(t,x,\cdot)|| \le K.$$

Then there exist: a collection $\kappa\in\mathcal{K}$, a homogeneous process $(\xi_t, \mathcal{F}_t^\kappa)$ with independent increments and the characteristic function (3.4), independent of z_t^κ and a process $x_t^{s,x}$ progressively measurable with respect to $(\mathcal{F}_t^\kappa)$ and satisfying the equation

$$x_t = x + \int_0^t b(s + r, x_r)dr + \int_0^t \int c(s + r, x_r, z)p^\kappa(drdz) + \delta\xi_t.$$

Fix an arbitrary $\kappa\in\mathcal{K}$. Let $\mathcal{A}^\kappa$ be a class consisting of all processes $\alpha : [0,T]\times\Omega^\kappa \to A$ progressively measurable with respect to $(\mathcal{F}_t^\kappa)$. By the assumptions I, II for every $\kappa\in\mathcal{K}$, $\alpha\in\mathcal{A}^\kappa$ the equation

$$x_t = x + \int_0^t b(\alpha_r, s + r, x_r)dr + \int_0^t \int c(\alpha_r, s + r, x_r, z)\, p^\kappa\,(drdz)$$

has a unique solution $x_t^{\kappa,d,s,x}$.

<u>Lemma 3.4.</u> Let the assumptions I, II, IV hold, $u\in\Lambda_{loc}$ and satisfies (1.1), f satisfies (1.1) uniformly with respect to α, $Fu \le 0$ a.e. H_T.

Then for every $\alpha\in\mathcal{A}^\kappa$, $\kappa\in\mathcal{K}$ the process

$$\int_0^t e^{-\varphi_r} f(\alpha_r, s + r, x_r)dr + e^{-\varphi_t} u(s + t, x_t)$$

is a supermartingale with respect to $(\mathcal{F}_t^\kappa, P^\kappa)$, where

$$x_t = x_t^{\kappa,\alpha,s,x}, \quad \varphi_t = \varphi_t^{\kappa,\alpha,s,x} = \int_0^t r(\alpha_u, s + u, x_u^{\kappa,\alpha,s,x})du.$$

<u>Proof.</u> Fix $\kappa\in\mathcal{K}$, $\beta\in A$ and open bounded set $Q\subset\bar{Q}\subset(0,T)\times R^d$. From the inequality $Fu \le 0$ a.e. H_T it follows that for sufficiently large n on Q

$$-L^\beta u^{(n)} \ge f(\beta) + \delta_n^\beta, \tag{3.8}$$

where $\delta_n^\beta = (L^\beta u)^{(n)} - L^\beta u^{(n)} + f(\beta)^{(n)} - f(\beta)$. It is not difficult to prove, that δ_n^β are uniformly bounded on Q and $\delta_n^\beta \to 0$ a.e. Q as $n \to \infty$.

Let τ be the first exit time of the process $(s + t, x_t^{\kappa,\beta,s,x})$ from Q. Applying Ito formula to $u^{(n)}$ and using (3.8) we obtain

$$u^{(n)}(s,x) \geq E_{s,x}^{\kappa,\beta}\{e^{-\varphi_\tau}u^{(n)}(s + \tau, x_\tau) + \\ + \int_0^\tau e^{-\varphi_r}[f(\beta,s + r, x_r) - \delta_n^\beta(s + r,x_r) \, dr\}.$$

From this inequality, applying Lemma 3.1 and tending $n \to \infty$ we derive that for almost all $(s,x)\in Q$

$$u(s,x) \geq E_{s,x}^{\kappa,\beta} [e^{-\varphi_\tau}u(s +\tau, x_\tau) + \int_0^\tau e^{-\varphi_r}f(\beta,s + r, x_r)dr].$$

As Q was chosen arbitrary, from this inequality it is easy to derive, that for almost all $(s,x)\in H_t$

$$u(s,x) \geq E_{s,x}^{\kappa,\beta} [e^{-\varphi_{t-s}}u(t,x_{t-s}) + \int_0^{t-s} e^{-\varphi_r}f(\beta,s + r, x_r \, dr].$$

Note that both sides of this inequality are continuous in (s,x). Consequently it holds for all $(s,x)\in H_t$. The rest part of the proof is analoguous to that of Lemma III. 3.5 [2].

<u>Lemma 3.5.</u> Let the assumptions I, II hold, $u\in\Lambda^+$, f satisfies (1.1) uniformly with respect to α, $Fu \geq 0$ a.e.H_T.

Then for every $(s,x)\in H_T$

$$u(s,x) \leq \sup_{\kappa\in\mathcal{K}} \sup_{\alpha\in\mathfrak{A}^\kappa} E_{s,x}^{\kappa,\alpha} [\int_0^{T-s} e^{-\varphi_t}f(\alpha_t,s + t,x_t)dt + e^{-\varphi_{T-s}}u(T,x_{T-s})].$$

<u>Proof.</u> Fix an arbitrary open bounded set $Q\subset\bar{Q}\subset(0,T)\times R^d$. Let $(s,x)\in Q$. Set $\tilde{f}^\alpha(t,x) = f(\alpha,t,x) + L^\alpha u''(t,x)$ on Q, $\tilde{f}^\alpha(t,x) = f(\alpha,t,x)$ on $H_T\setminus Q$.

Fix $\delta > 0$, $\gamma\in (1,2)$ and define on operator $\tilde{F}_\delta$ by the formula
$$\tilde{F}_\delta u'(t,x) = \sup_{\alpha\in A}[L^\alpha u' + \tilde{f}^\alpha](t,x) + C_\gamma\delta^\gamma \int_{|y|\leq 1} \Delta_y^2 \, u'(t,x) \frac{dy}{|y|^{d+\gamma}} .$$

By the definition of u' and the inequality $Fu \geq 0$ a.e. H_T we have $\tilde{F}_\delta u' \geq 0$ a.e. Q.

Fix an arbitrary $\varepsilon > 0$. It is not difficult to prove, that there exist a Borel function $\alpha_\varepsilon : H_T \to A$ such that $\tilde{F}_\delta^{\alpha_\varepsilon} \, u' \geq -\varepsilon$ a.e. Q.

Denote $\delta_n^\varepsilon = (L^{\alpha_\varepsilon} u')^{(n)} - L^{\alpha_\varepsilon}(u')^{(n)} + (\tilde{f}^{\alpha_\varepsilon})^{(n)} - \tilde{f}^{\alpha_\varepsilon}$. It is easy to prove, that δ_n^ε are bounded on Q uniformly with respect to n and $\delta_n^\varepsilon \to 0$ a.e. Q as $n \to \infty$.

By Lemma 3.3 there exist a collection $\kappa \in \mathcal{K}$ and a process $(\xi_t, \mathcal{F}_t^\kappa)$ such that the equation

$$x_t = x + \int_0^t b(\alpha_\varepsilon(s + r, x_r), s + r, x_r)dr +$$

$$+ \int_0^t \int c(\alpha_\varepsilon(s + r, x_r), s + r, x_r, z)p^\kappa(drdz) + \delta\xi_t.$$

has a solution $x_t^{s,x}(\varepsilon,\delta)$. Applying Ito formula to $(u')^{(n)}$ we derive

$$(3.9)\qquad \begin{aligned}(u')^{(n)}(s,x) \le E_{s,x}^\kappa \{e^{-\varphi_\tau^\varepsilon}(u')^{(n)}(s + \tau, x_\tau(\varepsilon,\delta)) +\\ + \int_0^\tau e^{-\varphi_t^\varepsilon}[\tilde{f}^{\alpha_\varepsilon}(s + t, x_t(\varepsilon,\delta)) + \delta_n^\varepsilon(s + t, x_t(\varepsilon,\delta)) + \varepsilon]dt\},\end{aligned}$$

where τ is the first exit time of the process $(s + t, x_t^{s,x}(\varepsilon,\delta))$ from Q,

$$\varphi_t = \int_0^t r(\alpha_\varepsilon\ (s + \rho, x_\rho^{s,x}(\varepsilon,\delta)), s + \rho, x_\rho^{s,x}(\varepsilon,\delta))d\rho.$$

Tending $n \to \infty$ in (3.9) and using Lemma 3.2 we obtain

$$(3.10)\qquad \begin{aligned}u'(s,x) \le E_{s,x}^\kappa \{e^{-\varphi_\tau^\varepsilon} u'(s + \tau, x_\tau(\varepsilon,\delta)) +\\ + \int_0^\tau e^{-\varphi_t^\varepsilon}(\tilde{f}^{\alpha_\varepsilon}(s + t, x_t(\varepsilon,\delta)) + \varepsilon]dt\}.\end{aligned}$$

Applying Ito formula to $(u")^{(n)}$, tending $n \to \infty$ and using Lemma 3.2 we derive

$$(3.11)\qquad \begin{aligned}u"(s,x) = E_{s,x}^\kappa \{e^{-\varphi_\tau^\varepsilon} u"(s + \tau, x_\tau(\varepsilon,\delta)) +\\ + \int_0^\tau e^{-\varphi_t^\varepsilon}[L^{\alpha_\varepsilon}u" + C_\gamma\delta^\gamma L_\gamma u"](s + t, x_t(\varepsilon,\delta))dt\}\end{aligned}$$

By Lemma 3.2

$$(3.12)\qquad \delta^\gamma E_{s,x}^\kappa \int_0^\tau |L_\gamma u"|(s + t, x_t(\varepsilon,\delta))dt \to 0 \quad \text{as } \delta \to 0.$$

From (3.10)-(3.12) we derive

$$(3.13)\quad u(s,x) \le E^{\kappa}_{s,x}\{e^{-\psi^{\varepsilon}_{\tau}} u(s+\tau, x_{\tau}(\varepsilon,\delta)) + \int_0^{\tau} e^{-\psi^{\varepsilon}_t} f(\alpha_{\varepsilon}(s+t, x_t(\varepsilon,\delta)), s+t, x_t(\varepsilon,\delta))dt\} + h(\varepsilon,\delta),$$

where $h(\varepsilon,\delta) \to 0$ as $\delta \to 0$, $\varepsilon \to 0$.

Set $\beta_t = \alpha_{\varepsilon}(s+t, x^{s,x}_t(\varepsilon,\delta))$. By assumptions I, II the equation

$$y_t = x + \int_0^t b(\beta_r, s+r, y_r)dr + \int_0^t \int c(\beta_r, s+r, y_r, z)p^{\kappa}(drdz)$$

has the unique solution $y^{s,x}_t$. It is easy to prove that

$$(3.14)\quad E^{\kappa} \sup_{t\le T-s} | x^{s,x}_t(\varepsilon,\delta) - y^{s,x}_t|^2 \to 0 \quad \text{as} \quad \delta \to 0$$

Let Q_1 be an open set, such that $Q_1 \subset Q_1 \subset Q$, τ_1 the first exit time of the process $(s+t, y^{s,x}_t)$ from Q_1, $\sigma = \tau\wedge\tau_1$. Note that by (3.14)

$$P^{\kappa}(\sigma \ne \tau_1) = P^{\kappa}(\tau > \tau_1) \le P^{\kappa}(\sup_{t<T-s} |x^{s,x}_t(\varepsilon,\delta) - y^{s,x}_t| >$$
$$> \operatorname{dist}(\partial Q_1, \partial Q)) \to 0 \quad \text{as} \quad \delta \to 0.$$

From (3.13) and (3.15) using the assumptions of the lemma we derive

$$u(s,x) \le E^{\kappa}\{e^{-\psi_{\tau_1}} u(s+\tau_1, y^{s,x}_{\tau_1}) + \int_0^{\tau_1} e^{-\psi_t} f(\beta_t, s+t, y^{s,x}_t)dt\} + h(\varepsilon,\delta) + \tilde{h}(\delta),$$

where $\tilde{h}(\delta) \to 0$ as $\delta \to 0$. Since $\varepsilon > 0$, $\delta > 0$ were arbitrary we obtain

$$u(s,x) \le \sup_{\kappa\in\mathcal{K}} \sup_{\alpha\in\mathcal{A}^{\kappa}} E^{\kappa,\alpha}_{s,x}\{e^{-\psi_{\sigma}} u(s+\sigma, x_{\sigma}) + \int_0^{\sigma} e^{-\psi_t} f(\alpha_t, s+t, x_t)dt\}.$$

Since Q was an arbitrary open bounded set, from this inequality and well known estimates for the moments of $x^{\kappa,\alpha,s,x}_t$ it follows the statement of lemma.

Proof of Theorem 1.2. By Lemmas 3.4, 3.5 we have the following probabilistic representation of a solution of the equation Fu = 0 a.e. H_T

$$u(s,x) = \sup_{\kappa\in\mathcal{K}} \sup_{\alpha\in\mathfrak{A}^\kappa} E^{\kappa,\alpha}_{s,x} \{e^{-\varphi_{T-s}} u(T,x_{T-s}) + \int_0^{T-s} e^{-\varphi_t} f(\alpha_t, s+t, x_t)\, dt\}.$$

From this representation the statement of theorem follows immediately.

References

[1] N.V. Krylov, Some new results from the theory of controlled diffusion processes, Matem. sbornik, 109 (151), No 1 (1979), 146-164 (in Russian).

[2] N.V. Krylov, Controlled diffusion processes, "Nauka", Moscow, 1977 (in Russian).

[3] N.V. Krylov, H. Pragarauskas, On traditional derivation of Bellman equation for general controlled stochastic processes, Liet. matem. rink., XXI, No 2 (1981), 101-110 (in Russian).

[4] S. Anulova, H. Pragarauskas, On Markov weak solutions of stochastic equations, Liet matem. rink., XVII, No 2 (1977), 5-26 (in Russian).

Institute of Mathematics and Cybernetics
Academy of Sciences of the Lithuanian SSR
Vilnius, K. Pozelos 54

A LIMIT THEOREM OF SOLUTIONS OF STOCHASTIC BOUNDARY-INITIAL-VALUE PROBLEMS

Jürgen vom Scheidt (Zwickau)

1. Introduction

This paper deals with stochastic partial differential equation problems. The inhomogeneous term of a differential equation is assumed to be a random function $f(x,\omega)$. This random function is a so-called weakly correlated function, i.e. we consider random functions which do not possess a "distance effect". At a fixed x the domain of dependence of the weakly correlated function $f(x,\omega)$ is characterized by a parameter ε. In the present paper limit theorems for the solutions of differential equation problems are proved as $\varepsilon \downarrow 0$. These limit theorems are of the central limit theorem type. It is proved that the solution converges in distribution to a Gaussian random function. Thus, we will be able to deal with initial-value problems, boundary-value problems, and eigenvalue problems with random coefficients in the same way ([2], [3]).

We will apply the results to the calculation of random vibrations of beams and plates with an external random excitation. The probability distribution functions of displacements and stresses can be calculated approximately.

2. Weakly correlated random functions

Definition 1. A random function $f_\varepsilon(x,\omega)$, $x \in D \subset R^m$ where $E\{f_\varepsilon(x)\} \doteq \langle f_\varepsilon(x) \rangle = 0$ is called weakly correlated of the correlation length ε if

$$\langle f_\varepsilon(\overset{1}{x}) f_\varepsilon(\overset{2}{x}) \dots f_\varepsilon(\overset{k}{x}) \rangle$$

$$= \langle f_\varepsilon(\overset{11}{x}) \dots f_\varepsilon(\overset{1k_1}{x}) \rangle \langle f_\varepsilon(\overset{21}{x}) \dots f_\varepsilon(\overset{2k_2}{x}) \rangle \dots \langle f_\varepsilon(\overset{p1}{x}) \dots f_\varepsilon(\overset{pk_p}{x}) \rangle$$

for the k-th moments (for all $k \geq 1$). In the above, $\{(\overset{11}{x}, \dots, \overset{1k_1}{x}), (\overset{21}{x}, \dots, \overset{2k_2}{x}), \dots, (\overset{p1}{x}, \dots, \overset{pk_p}{x})\}$ denotes the splitting of $(\overset{1}{x}, \dots, \overset{k}{x})$ in the maximum ε-adjoining subsets. A subset $(\overset{i1}{x}, \dots, \overset{il}{x}) \subset (\overset{1}{x}, \dots, \overset{k}{x})$ is said to be maximum ε-adjoining if it is ε-adjoining i.e. if a permutation $(\overset{i1}{y}, \dots, \overset{il}{y})$ of $(\overset{i1}{x}, \dots, \overset{il}{x})$ exists where $|\overset{is}{y} - \overset{is+1}{y}| \leq \varepsilon$,

$s=1,2,\ldots,l-1$ $(|x|^2 \doteq \sum_{i=1}^{m} x_i^2)$, but $(\overset{s}{x},\overset{i_1}{x},\ldots,\overset{i_l}{x})$ isnot ε-adjoining for a $\overset{s}{x} \in (\overset{1}{x},\ldots,\overset{k}{x}) \setminus (\overset{i_1}{x},\ldots,\overset{i_l}{x})$. $(\overset{i}{x})$ is defined to be ε-adjoining.

The idea concerning the definition of a weakly correlated function can also be found in papers of Ornstein, Uhlenbeck [4] and Boyce [1].

The limit theorems for solutions of random differential equation problems are based upon a limit theorem for linear functionals of the type

$$r_{i\varepsilon}(x,\omega) \doteq \frac{1}{\sqrt{\varepsilon}^m} \int_D F_i(x,y) f_\varepsilon(y,\omega)\, dy \quad i=1,2,\ldots,l$$

where D denotes a bounded domain and $F_i(x,y)$ functions on $G \times D \subset R^p \times R^m$. Let $F_i(x,y)$ be from $L_2(D)_y$ for each $x \in G$. Finally, we state without proof a result for these functionals.

<u>Theorem 1.</u> Let $\{f_\varepsilon(x,\omega)\}_{\varepsilon \downarrow 0}$ be a sequence of weakly correlated functions with continuous sample functions and $\langle |f(x)|^p \rangle \le d_p < \infty$, $p=1,2,\ldots$ Then we get in distribution

$$\lim_{\varepsilon \downarrow 0} \frac{1}{\sqrt{\varepsilon}^m} \{r_{1\varepsilon}(x,\omega), r_{2\varepsilon}(x,\omega),\ldots,r_{l\varepsilon}(x,\omega)\} = \{\xi_1(x,\omega),\ldots,\xi_l(x,\omega)\}$$

where $\xi(x,\omega) = \{\xi_1(x,\omega),\ldots,\xi_l(x,\omega)\}$ is a Gaussian vector function. $\xi(x,\omega)$ has the moments $\langle \xi_i(x) \rangle \equiv 0$, $i=1,2,\ldots l$, and

$$\langle \xi_i(\overset{1}{x}) \xi_j(\overset{2}{x}) \rangle = \int_D F_i(\overset{1}{x},y) F_j(\overset{2}{x},y) a(y)\, dy, \quad i,\ j=1,2,\ldots,l$$

$a(y)$ denotes a characteristic function relative to the random function $f_\varepsilon(x,\omega)$ and is defined by

$$a(y) = \lim_{\varepsilon \downarrow 0} \frac{1}{\varepsilon^m} \int_{K_\varepsilon(0)} \langle f_\varepsilon(y) f_\varepsilon(y+z) \rangle dz$$

where $K_\varepsilon(y) = \{z \in R^m : |y-z| \le \varepsilon\}$. $a(y)$ is called the intensity of f_ε.

<u>Remark</u>. 1. Let $f_\varepsilon(x,\omega)$ be a wide-sense homogeneous function.

Then we have $a(y) \equiv a = \lim_{\varepsilon \downarrow 0} \frac{1}{\varepsilon^m} \int_{K_\varepsilon(0)} \langle f_\varepsilon(0) f_\varepsilon(z) \rangle dz$.

2. The above theorem can be expanded to the convergence of the moments $\langle r_{i_1\varepsilon}(\overset{1}{x}) \ldots r_{i_p\varepsilon}(\overset{p}{x}) \rangle$ to the moments $\langle \xi_{i_1}(\overset{1}{x}) \ldots \xi_{i_p}(\overset{p}{x}) \rangle$.

The proof of the theorem 1 we can find in [5].

A random function can have the property that it is weakly correlated relative to a group of independent variables. For instance, it is possible that $f_\varepsilon(t,x,\omega)$ is weakly correlated relative to t or x. Hence, we give the definition 2.

Definition 2. A random function $f_\varepsilon(x,y,\omega)$, $(x,y)\in R^s x R^t$, $s+t=m$, with $<f_\varepsilon(x,y)>\equiv 0$ is said to be partially weakly correlated relative to x and the correlation length ε if

$$<\prod_{i=1}^{k} f_\varepsilon(\overset{i}{x},\overset{i}{y})> = \prod_{s=1}^{p} < \prod_{j=1}^{k_s} f_\varepsilon(\overset{sj}{x},\overset{sj}{y})>$$

for all $k\geq 1$ $(\sum_{i=1}^{p} k_i = k)$. $\{(\overset{11}{x},\ldots,\overset{1k_1}{x}),(\overset{21}{x},\ldots,\overset{2k_2}{x}),\ldots,(\overset{p1}{x},\ldots,\overset{pk_p}{x})\}$

denotes the splitting of $(\overset{1}{x},\ldots,\overset{k}{x})$ in the maximum ε-adjoining subsets of R^s and $\overset{i}{y}$, $i=1,2,\ldots,k$, are arbitrary points in R^t. A simple example of a partially weakly correlated function is $f_\varepsilon(x,y,\omega) = f_{1\varepsilon}(x,\omega)f_2(y)$ where $f_{1\varepsilon}(x,\omega)$ denotes a weakly correlated function and $f_2(y)$ a deterministic function.

$$a(x;\overset{1}{y},\overset{2}{y}) \doteq \lim_{\varepsilon\downarrow 0} \frac{1}{\varepsilon^s} \int_{K_\varepsilon(0)} <f_\varepsilon(x,\overset{1}{y})f_\varepsilon(x+z,\overset{2}{y})>dz$$

where $K_\varepsilon(0)\subset R^s$. For the above simple example we get $a(x;\overset{1}{y},\overset{2}{y}) =$

$a(x)f_2(\overset{1}{y})f_2(\overset{2}{y})$ where $a(x) = \lim_{\varepsilon\downarrow 0} \frac{1}{\varepsilon^s} \int_{K_\varepsilon(0)} <f_{1\varepsilon}(x)f_{1\varepsilon}(x+z)>dz.$

Thus, we obtain .

Theorem 2. Let $\{f_\varepsilon(y,z,\omega)\}_{\varepsilon\downarrow 0}$ be a sequence of partially weakly correlated functions relative to y with continuous sample functions and $<|f(y,z)|^p> \leq d_p < \infty$ for $p=1,2,\ldots$ It is $x\in G\subset R^p$ and $D = D_1 x D_2\subset R^s x R^t$ where D_1, D_2 denote bounded domains. Let $F(x;y,z)$ be a function defined on $G x D_1 x D_2$ and $F(x;y,z)\in L_2(D_1)_y$. Then

$$r_\varepsilon(x,\omega) = \frac{1}{\sqrt{\varepsilon^s}} \int_D F(x;y,z)\; f_\varepsilon(y,z,\omega)dydz$$

converges in distribution to the Gaussian random function $\xi(x,\omega)$,

$$\lim_{\varepsilon\downarrow 0} r_\varepsilon(x,\omega) = \xi(x,\omega) \quad \text{in distribution.}$$

The moments of $\xi(x,\omega)$ can be written as $<\xi(x)> = 0$ and

$$<\xi(\overset{1}{x})\xi(\overset{2}{x})> = \int_{D_2}\int_{D_2}\int_{D_1} F(\overset{1}{x};y,\overset{1}{z})F(\overset{2}{x};y,\overset{2}{z})a(y;\overset{1}{z},\overset{2}{z})dyd\overset{2}{z}d\overset{1}{z}.$$

In this case we also have the remark 2 of theorem 1. The intensity $a(y;\overset{1}{z},\overset{2}{z})$ is independent of y if the function $f_\varepsilon(y,z,\omega)$ is a homogeneous function relative to y. The proof of theorem 2 can be obtained from theorem 1.

Finally, we give the definition of weakly correlated connected random functions.

Definition 3. A random vector function $\{f_{1\varepsilon}(x,\omega),f_{2\varepsilon}(x,\omega)\}$, $x\in D\subset R^m$, with $\langle f_{1\varepsilon}(x)\rangle\equiv\langle f_{2\varepsilon}(x)\rangle\equiv 0$ is said to be weakly correlated connected of the correlation length if

$$\langle \prod_{i=1}^{k} f_{r_i\varepsilon}(\overset{i}{x})\rangle = \prod_{s=1}^{p} \langle \prod_{j=1}^{k_s} f_{r_{sj}\varepsilon}(\overset{sj}{x})\rangle \quad (\sum_{s=1}^{p} k_s = k)$$

for all $k\geq 1$. $\{(\overset{11}{x},\ldots,\overset{1k_1}{x}),\ldots,(\overset{p1}{x},\ldots,\overset{pk_p}{x})\}$ denotes the splitting of $(\overset{1}{x},\ldots,\overset{k}{x})$ in the maximum ε-adjoining subsets of R^m and $r_i = 1,2$. The definition of a weakly correlated connected vector function implies weak correlation of components of the vector function. It follows from the independence of the weakly correlated functions $f_{1\varepsilon}(x,\omega)$, $f_{2\varepsilon}(x,\omega)$ that the vector function $\{f_{1\varepsilon},f_{2\varepsilon}\}$ is weakly correlated connected.

Theorem 3. Let $\{f_{1\varepsilon}(x,\omega),f_{2\varepsilon}(x,\omega)\}_{\varepsilon\downarrow 0}$ be a sequence of weakly correlated connected vector functions on $D\subset R^m$ where $f_{i\varepsilon}(x,\omega)$, $i=1,2$, have continuous sample functions and $\langle |f_{i\varepsilon}(x)|^p\rangle \leq d_p < \infty$, $i=1,2$, $p=1,2,\ldots$ We consider a linear functional

$$(1)\quad r_\varepsilon(x,\omega) = \frac{1}{\sqrt{\varepsilon^m}} \int_D F_1(x,y)f_{1\varepsilon}(y,\omega)dy + \frac{1}{\sqrt{\varepsilon^m}} \int_D F_2(x,y)f_{2\varepsilon}(y,\omega)dy.$$

The functions $F_1(x,y)$, $F_2(x,y)$ are defined on GxD and $F_i(x,y)\in L_2(D)_y$. Then the sequence of random functions $\{r_\varepsilon(x,\omega)\}_{\varepsilon\downarrow 0}$ satisfies

$$\lim_{\varepsilon\downarrow 0} r_\varepsilon(x,\omega) = \xi(x,\omega) \qquad \text{in distribution},$$

where $\xi(x,\omega)$ is a Gaussian random function possessing the moments $\langle\xi(x)\rangle\equiv 0$ and

$$\langle\xi(\overset{1}{x})\xi(\overset{2}{x})\rangle = \sum_{i,j=1}^{2} \int_D F_i(\overset{1}{x},y)F_j(\overset{2}{x},y)a_{ij}(y)dy.$$

The intensity $a_{ij}(y)$ of the random functions $f_{i\varepsilon}$ and $f_{j\varepsilon}$ is defined by

$$a_{ij}(y) \doteq \lim_{\varepsilon \downarrow 0} \frac{1}{\varepsilon^m} \int_{K_\varepsilon(0)} \langle f_{i\varepsilon}(y) f_{j\varepsilon}(y+z) \rangle dz, \quad i,j=1,2.$$

A generalization of this theorem to random vector functions $\{r_{1\varepsilon}(x,\omega),\dots,r_{1\varepsilon}(x,\omega)\}$ can be carried out in a straightforward fashion. Theorem 3 can also be proved for partially weakly correlated connected vector functions. It follows from the independence of $f_{1\varepsilon}$ and $f_{2\varepsilon}$ that $a_{12}(y) \equiv a_{21}(y) \equiv 0$. In this case we can write down the random function $\xi(x,\omega)$ as a sum of two independent Gaussian functions $\xi_1(x,\omega)$, $\xi_2(x,\omega)$ where

$$\langle \xi_i(\overset{1}{x}) \xi_i(\overset{2}{x}) \rangle = \int_D F_i(\overset{1}{x},y) F_i(\overset{2}{x},y) a_{ii}(y) dy.$$

3. Boundary-initial-value problems

We consider the boundary-initial-value problem on $[0,T] \times D$

$$\frac{\partial^2 u}{\partial t^2} + 2\beta \frac{\partial u}{\partial t} + Lu = p(t,x,\omega) \tag{2}$$

with the initial conditions $u(0,x) = u_o(x)$, $\frac{\partial u}{\partial t}(0,x) = u_1(x)$ and the boundary conditions on ∂D according to the order of the elliptic differential operator L. Let D be a bounded domain of R^n. Many problems of vibrating damped beams and plates under stochastic loading lead to the boundary-initial-value problem (2).

Using Fourier method the formal solution of (2) can be found easily. Let μ_i and $w_i(x)$ be the eigenvalues and the orthonormal eigenfunctions of L and the boundary conditions. We obtain

$$u(t,x) = \sum_{i=1}^{\infty} [(a_i \cos \beta_i t + b_i \sin \beta_i t) e^{-\beta t} + \int_0^t h_i(t-s) P_i(s) ds] w_i(x) \tag{3}$$

where $\beta_i \doteq (\mu_i - \beta^2)^{1/2}$, $h_i(t) = \frac{1}{\beta_i} e^{-\beta t} \sin \beta_i t$, $P_i(t) = (p(t,.), w_i)$ and a_i, b_i can be determined from

$$u_o(x) = \sum_{i=1}^{\infty} a_i w_i(x), \quad u_1(x) = \sum_{i=1}^{\infty} (\beta_i b_i - \beta a_i) w_i(x)$$

by $a_i = (u_o, w_i)$, $b_i = \frac{1}{\beta_i} (\beta a_i + (u_1, w_i))$.

Substituting $\langle p(t,x) \rangle$ into (2) we obtain the averaged problem. The solution of this averaged problem is denoted by $w(t,x)$ and it is

$\overline{u}(t,x,\omega) \doteq u(t,x,\omega)-w(t,x)$. We have from Eq. (3) (boundary- and initial conditions are deterministic)

$$\overline{u}(t,x,\omega) = \sum_{i=1}^{\infty} \int_{o}^{t} h_i(t-s)\overline{P}_i(s)\,ds\,w_i(x)$$

where $\overline{P}_i(t) = (\overline{p}(t,.),w_i)$, $\overline{p}(t,x,\omega) \doteq p(t,x,\omega)-\langle p(t,x)\rangle$. The random function $\overline{u}(t,x,\omega)$ is a weak solution a.s. of the differential equation

$$\frac{\partial^2\overline{u}}{\partial t^2} + 2\,\frac{\partial\overline{u}}{\partial t} + L\overline{u} = \overline{p}(t,x,\omega) \tag{4}$$

with homogeneous initial and boundary conditions where $p(t,x,\omega)$ has smooth sample functions a.s. That means that

$$(\overline{u}_{tt},g) + 2\beta(\overline{u}_t,g) + (\overline{u},Lg) = (\overline{p},g) \quad \text{a.s.}$$

is fulfilled for all $g\in C_o^\infty(D)$. The derivatives $\overline{u}_t$, $\overline{u}_{tt}$ denote derivatives in $L_2(D)$.

Now we consider the "averaged" solution of the boundary-initial-value problem (4)

$$\overline{U}(t,\omega) \doteq (\overline{u}(t,.),g) = \sum_{i=1}^{\infty} \int_{o}^{t} h_i(t-s)(w_i,\overline{p}(s,.))\,ds\,(w_i,g)$$

where $g\in C_o^\infty(D)$. Using

$$F(t,y) \doteq \sum_{i=1}^{\infty} h_i(t)w_i(y)(w_i,g)$$

it is

$$\overline{U}(t,\omega) = \int_{o}^{t}\int_{D} F(t-s,y)\overline{p}(s,y)\,dy\,ds$$

because the series of $F(t,y)$ possesses good properties of convergence. Let

$$F(t,y,x) \doteq \sum_{i=1}^{\infty} h_i(t)w_i(y)w_i(x)$$

be and $F(t,y,x)\in L_2([0,T]xDxD)$. Then we have

$$\overline{U}(t,\omega) = (\int_{o}^{t}\int_{D} F(t-s,y,x)\overline{p}(s,y)\,dy\,ds,g(x)).$$

Assuming the above condition and the uniform convergence of $\sum_{i=1}^{\infty} h_i^2(t)w_i^2(x)$ relative to x then $\int_{o}^{t}\int_{D} F(t-s,y,x)\overline{p}(s,y)\,dy\,ds$ is a continuous function of x. The solution of Eq. (4) can be written by

$$\overline{u}(t,x) = \int_0^t \int_D F(t-s,y,x)\overline{p}(s,y)dyds. \tag{5}$$

Furthermore, using $D_t D_x^{\ell}\overline{u} \doteq \frac{\partial^{\ell+1}\overline{u}}{\partial t \partial x_{i_1}^{\ell_1} \ldots \partial x_{i_p}^{\ell_p}}$ $\quad (\sum_{s=1}^{p} \ell_s = \ell)$,

we obtain

$$(D_t D_x^{\ell}\overline{u}(t,x,\omega),g(x)) = \int_0^t \int_D D_t D_x^{\ell} F(t-s,y)\overline{p}(s,y)dyds$$

where $D_t D_x^{\ell} F(t,y) = \sum_{i=1}^{\infty} \frac{\partial h_i(t)}{\partial t} w_i(y)(D_x^{\ell} w_i,g)$. It is

$$(D_t D_x^{\ell}\overline{u}(t,x,\omega),g(x)) = (\int_0^t \int_D D_x^{\ell} F(t-s,y,x)\overline{p}(s,y)dyds,g(x))$$

if we make use of

$$D_t D_x^{\ell} F(t,y,x) \doteq \sum_{i=1}^{\infty} \frac{\partial h_i}{\partial t}(t) w_i(y) D_x^{\ell} w_i(x) \in L_2([0,T]xDxD).$$

The equation

$$D_t D_x^{\ell}\overline{u}(t,x,\omega) = \int_0^t \int_D D_t D_x^{\ell} F(t-s,y,x)\overline{p}(s,y,\omega)dyds \tag{6}$$

follows from similar considerations as for $\overline{u}(t,x,\omega)$ having the continuity of $\int_0^t \int_D D_t D_x^{\ell} F(t-s,y,x)\overline{p}(s,y)dyds$ relative to x.

3.1. Weakly correlated inhomogeneous parts

Theorem 1 can be applied to the solution (5) and (6). Assuming, that $\overline{p}(t,x,\omega)$ is a weakly correlated function $\overline{p}_{\varepsilon}(t,x,\omega)$ and $D_x^{\ell} F(t,y,x) \in L_2(D)_y$ we have

$$\lim_{\varepsilon \downarrow 0} \frac{1}{\sqrt{\varepsilon}^{n+1}} D_x^{\ell}\overline{u}(t,x,\omega) = \eta_{\ell}(t,x,\omega) \text{ in distribution}$$

where $\langle \eta_{\ell}(t,x) \rangle \equiv 0$ and

$$\langle \eta_\ell(t_1,x)\eta_\ell(t_2,x)\rangle =$$

(7)

$$= \int_0^{\min(t_1,t_2)} \int_D D^\ell_{\overset{1}{x}} F(t_1-s,y,x) D^\ell_{\overset{2}{x}} F(t_2-s,y,x)\ a(s,y)dyds.$$

That means that $D^\ell_x u(t,x,\omega)$ denotes for a small ε approximately a Gaussian random function with the moments

$$\langle D^\ell_x u(t,x)\rangle \simeq D^\ell_x w(t,x), \langle D^\ell_{\overset{1}{x}} u(t_1,\overset{1}{x}) D^\ell_{\overset{2}{x}} u(t_2,\overset{2}{x})\rangle \simeq \varepsilon^{n+1} \langle \eta_\ell(t_1,\overset{1}{x})\eta_\ell(t_2,\overset{2}{x})\rangle$$

If $\overline{p}_\varepsilon(t,x,\omega)$ is a wide-sense homogeneous function we know that $a(t,x)\equiv a\equiv\text{const}$. From Eq. (7) we find

$$\langle \eta_o(t_1,\overset{1}{x})\eta_o(t_2,\overset{2}{x})\rangle = a \sum_{i=1}^{\infty} T_i(t_1,t_2) w_i(\overset{1}{x}) w_i(\overset{2}{x})$$

where

$$T_i(t_1,t_2) = \int_0^{\min(t_1,t_2)} h_i(t_1-s)h_i(t_2-s)ds.$$

As $t_1,t_2\to\infty$ with $t_2-t_1=s=\text{const}$ we have

$$\lim_{\substack{t_1,t_2\to\infty \\ -t_1+t_2=s}} T_i(t_1,t_2) = \frac{1}{4\mu_i} e^{-\beta|s|} \{\frac{1}{\beta} \cos\beta_i|s| + \frac{1}{\beta_i} \sin\beta_i|s|\} \doteq T_{io}(s).$$

Hence, the random function $\eta_\ell(t,x,\omega)$ denotes as $t_1,t_2\to\infty$ a stationary Gaussian function $\tilde{\eta}_\ell(t,x,\omega)$ relative to t where

$$\langle \tilde{\eta}_\ell(t_1,\overset{1}{x})\tilde{\eta}_\ell(t_2,\overset{2}{x})\rangle = \frac{a}{4} \sum_{i=1}^{\infty} T_{io}(s) w_i(\overset{1}{x}) w_i(\overset{2}{x}).$$

Example. These results can be applied to the motion of a string with constant density ρ of length 1, sourrounded by a gas ([6]).

The equation of motion of the string is

$$\frac{\partial^2 u}{\partial t^2} + f \frac{\partial u}{\partial t} - \frac{\partial}{\partial x} (\tau \frac{\partial u}{\partial x}) = F(t,x,\omega), \quad 0 \le x \le 1$$

where f is the friction coefficient, $\tau(x)$ the tension of the string and $F(t,x,\omega)$ is the fluctuating force. The boundary conditions are

$h_o u(t,0) - \frac{\partial u}{\partial x}(t,0) = 0$, $-h_\ell u(t,\ell) - \frac{\partial u}{\partial x}(t,\ell) = 0$ (string is bound elastically) and the initial conditions $u(0,x) = \frac{\partial u}{\partial t}(0,x) = 0$. We can calculate the intensity a of a homogeneous random function $F(t,x,\omega)$ from the theorem of the equipartion of energy

$$a = \frac{2fkT}{2}$$

where T is the absolute temperature and $k = \frac{R}{N}$ with the Loschmidt-number N and R the gas-constant. The above made results applied to this problem lead to numerical calculations of the limit function of $u(t,x,\omega)$.

3.2. Partially weakly correlated inhomogeneous parts

In this section let $\overline{p}_\varepsilon(t,x,\omega)$ be a partially weakly correlated function relative to t. A partially weak correlation relative to x can also be considered and we can obtain interesting results. In this theory we need the weakly correlated connected functions.

From section 2

$$(8)\quad D_t^k D_x^\ell \overline{u}(t,x,\omega) = \int_0^t \sum_{i=1}^\infty D_t^k h_i(t-s) D_x^\ell w_i(x)(w_i,\overline{p}(s,.))ds$$

is obtained if the right-hand side of (8) is a continuous function. The uniform convergence of series (8) relative to x implies this continuity. The equation $(w_i, p(t,.)) = \frac{1}{\mu_i^r}(w_i, L^r\overline{p}_\varepsilon(t,.))$, $r \geq 1$, leads to the uniform convergence where $\overline{p}_\varepsilon(t,x,\omega)$ must fulfil some boundary conditions. Therefore, we have for some r

$$D_t^k D_x^\ell \overline{u}(t,x,\omega) = \int_D \int_0^t \tilde{F}_{k\ell}^r(t-s,x,y) L^r\overline{p}_\varepsilon(s,y,\omega)dsdy$$

where $\tilde{F}_{k\ell}^r(t,x,y) \doteq \sum_{i=1}^\infty \frac{1}{\mu_i^r} D_t^k h_i(t) D_x^\ell w_i(x) w_i(y)$. Let $\overline{p}_\varepsilon(t,x,\omega)$ be a weakly correlated random function relative to t where

$$a(t;\overset{1}{x},\overset{2}{x}) = \lim_{\varepsilon\downarrow 0} \frac{1}{\varepsilon} \int_{-\varepsilon}^{\varepsilon} \langle \overline{p}_\varepsilon(t,\overset{1}{x})\overline{p}_\varepsilon(t+s,\overset{2}{x})\rangle ds$$

and

$$L_{\overset{1}{x}}^r L_{\overset{2}{x}}^r a(t;\overset{1}{x},\overset{2}{x}) = \lim_{\varepsilon\downarrow 0} \frac{1}{\varepsilon} \int_{-\varepsilon}^{\varepsilon} \langle L_{\overset{1}{x}}^r\overline{p}_\varepsilon(t,\overset{1}{x}) L_{\overset{2}{x}}^r\overline{p}_\varepsilon(t+s,\overset{2}{x})\rangle ds.$$

According to theorem 2,

$$\lim_{\varepsilon\downarrow 0} \frac{1}{\sqrt{\varepsilon}} \; D_t^k D_x^{\ell} \overline{u}(t,x,\omega) = \zeta_{k\ell}(t,x,\omega) \quad \text{in distribution holds.}$$

The Gaussian function $\zeta_{k\ell}(t,x,\omega)$ possesses the moments $\langle\zeta_{k\ell}(t,x)\rangle \equiv 0$ and

(9) $\langle\zeta_{k\ell}(t_1,\overset{1}{x})\zeta_{k\ell}(t_2,\overset{2}{x})\rangle =$

$$= \int_D \int_D \int_0^{\min(t_1,t_2)} \tilde{F}_k^r(t_1-s,\overset{1}{x},\overset{1}{y})\tilde{F}_{k\ell}^r(t_2-s,\overset{2}{x},\overset{2}{y}) L_{\overset{1}{y}}^r L_{\overset{2}{y}}^r a(s;\overset{1}{y},\overset{2}{y})\, ds\, d\overset{2}{y}\, d\overset{1}{y}$$

In the special case $\overline{p}_\varepsilon(t,x,\omega) = \overline{p}_{o\varepsilon}(t,\omega)p_1(x)$ we have

$$D_t^k D_x^{\ell}\overline{u}(t,x,\omega) = \int_0^t G_{k\ell}(t-s,x)\overline{p}_{o\varepsilon}(s,\omega)\,ds$$

with $G_{k\ell}(t,x) = \sum_{i=1}^{\infty} D_t^k h_{i_t}(t) D_x^{\ell} w_i(x)(w_i,p_1)$

from (8) where $G_{k\ell}(t,x)\in L_2([0,t]\times D)$ and $\int_0^t G_{k\ell}(t-s,x)\overline{p}_{o\varepsilon}(s,\omega)\,ds$ is continuous relative to x. The last condition is fulfilled if $G_{k\ell}(s,x)$ is continuous in $L_2([0,t])$ relative to x. From theorem 1 we obtain

$$\lim_{\varepsilon\downarrow 0} \frac{1}{\sqrt{\varepsilon}} \; D_t^k D_x^{\ell}\overline{u}(t,x,\omega) = \zeta_{k\ell}(t,x,\omega) \text{ in distribution}$$

where

$$\langle\zeta_{k\ell}(t_1,\overset{1}{x})\zeta_{k\ell}(t_2,\overset{2}{x})\rangle = \int_0^{\min(t_1,t_2)} G_{k\ell}(t_1-s,\overset{1}{x})G_{k\ell}(t_2-s,\overset{2}{x})a(s)\,ds$$

and the intensity to $\overline{p}_{o\varepsilon}(t,\omega)$ is a(t). These statements can be transferred to random vector functions $\{D_t^{k_1}D_x^{\ell_1}\overline{u}(t,x,\omega),\ldots, D_t^{k_p}D_x^{\ell_p}\overline{u}(t,x,\omega)\}$, e.g. to $\{\overline{u}_t(t,x)\cdot\overline{u}_t(t,x)\}$ or $\{\overline{u}_{x_ix_i}(t,x),\; \overline{u}_{x_ix_i}(t,x)\}$.

Let $\overline{p}_\varepsilon(t,x,\omega)$ be a wide-sense stationary process relative to t, i.e. $\langle\overline{p}_\varepsilon(t_1,\overset{1}{x})\overline{p}_\varepsilon(t_2,\overset{2}{x})\rangle \equiv K_\varepsilon(|t_1-t_2|,\overset{1}{x},\overset{2}{x})$. Hence we have $a(t;\overset{1}{x},\overset{2}{x}) = a(\overset{1}{x},\overset{2}{x})$ and similarly to (9)

$\langle\zeta_{p\ell}(t_1,\overset{1}{x})\zeta_{q\ell}(t_2,\overset{2}{x})\rangle =$

$$= \iint_{DD} \sum_{i,j=1}^{\infty} \frac{1}{\mu_i^r\mu_i^r} T_{ij}^{pq}(t_1,\, t_2)\; D_{\overset{1}{x}}^{\ell} w_i(\overset{1}{x}) D_{\overset{2}{x}} w_j(\overset{2}{x}) w_i(\overset{1}{y}) w_j(\overset{2}{y}) L_{\overset{1}{y}}^r L_{\overset{2}{y}}^r a(\overset{1}{y},\overset{2}{y})\, d\overset{2}{y}\, d\overset{1}{y}.$$

The random function $\frac{1}{\sqrt{\varepsilon}}\{D_t^p D_x^{\ell}\overline{u}(t,x), D_t^q D_x^{\ell}\overline{u}(t,x)\}$ converges as $\varepsilon\downarrow 0$ to $\{\zeta_{p\ell}(t,x),\zeta_{q\ell}(t,x)\}$. For $t_1,t_2\to\infty$ we obtain

$$\lim_{\substack{t_1,t_2\to\infty\\ t_1<t_2,\, t_2-t_1 = s}} T_{ij}^{oo}(t,t) = \frac{e^{-\beta_s}}{8\beta^2(\mu_i+\mu_j)+(\mu_j-\mu_i)^2}\,[4\beta\cos\beta_j s + \frac{4\beta^2+\mu_i-\mu_j}{\beta_j}\sin\beta_j s] \doteq T_{ijo}^{oo}(s)$$

and similar terms for other $T_{ij}^{pq}(t_1,t_2)$. Hence $\{\zeta_{o\ell},\zeta_{1\ell}\}$ is approximately a stationary Gaussian vector process for a large t. This stationary Gaussian vector function we note by $\{\tilde{\zeta}_{o\ell},\tilde{\zeta}_{1\ell}\}$. Particularly, $<\tilde{\zeta}_{o\ell}(t,\overset{1}{x})\tilde{\zeta}_{1\ell}(t,\overset{2}{x})> = 0$ for $\ell=0,1,\ldots$ follows and then $\tilde{\zeta}_{o\ell}(t,\overset{1}{x})$ and $\tilde{\zeta}_{1\ell}(t,\overset{2}{x})$ are independent random variables.

The spectral density

$$S_{p\ell q\ell}(\alpha;\overset{1}{x},\overset{2}{x}) \doteq \frac{1}{2\Pi}\int_{-\infty}^{\infty} e^{-i\alpha s} <\tilde{\zeta}_{p\ell}(t,\overset{1}{x})\tilde{\zeta}_{q\ell}(t+s,\overset{2}{x})>ds$$

to $<\tilde{\zeta}_{p\ell}(t_1,\overset{1}{x})\tilde{\zeta}_{q\ell}(t_2,\overset{2}{x})>$ is calculated by

$$S_{o\ell o\ell}(\alpha;x,x) =$$

$$\int_D\int_D \sum_{v,\,t=1}^{\infty} \frac{1}{\mu_v^r\mu_v^r} U_{vt}(\alpha)D_x^\ell w_v(x)D_x^\ell w_t(x)w_v(\overset{1}{y})w_t(\overset{2}{y})L_{\overset{1}{y}}^r L_{\overset{2}{y}}^r a(\overset{1}{y},\overset{2}{y})d\overset{2}{y}d\overset{1}{y}$$

where

$$U_{vt}(\alpha) = W_{vt}(\alpha) + W_{tv}(\alpha),$$

$$W_{vt}(\alpha) = \frac{8\beta^2\mu_t+(\mu_v-\mu_t)(\mu_t-\alpha)^2}{2\Pi[8\beta^2(\mu_v+\mu_t)+(\mu_v-\mu_t)^2]\,[(\mu_t+\alpha^2)^2-4\beta_t^2\alpha^2]}.$$

It is $S_{1\ell 1\ell}(\alpha;x,x) = \alpha^2 S_{o\ell o\ell}(\alpha;x,x)$. If $\overline{p}_\varepsilon(t,x,\omega) = \overline{p}_{o\varepsilon}(t,\omega)p_1(x)$, then we have

$$S_{o\ell o\ell}(\alpha;x,x) = 2a \sum_{v,t=1}^{\infty} W_{vt}(\alpha)D_x^\ell w_v(x)D_x^\ell w_t(x)(w_v,p_1)(w_t,p_1),$$

where the intensity to the wide-sense stationary process $\overline{p}_{o\varepsilon}(t,\omega)$ is a constant a.

<u>Example</u>. We consider the vibration of a simple supported Bernoulli-Euler beam

$$u_{tt}+2\beta u_t+Au_{xxxx} = p_{o\varepsilon}(t,\omega)p_1(x) \qquad 0 \le x \le 1$$

$$u(0,x) = u_t(0,x) = 0; \quad u(t,0) = u(t,\ell) = u_{xx}(t,0) = u_{xx}(t,\ell) = 0.$$

Let $p_{o\varepsilon}(t,\omega)$ be a weakly correlated and wide-sense stationary process. The intensity to $\overline{p}_{o\varepsilon}(t,\omega)$ is a. We have $\mu_i = A\ (\frac{i\Pi}{\ell})4$ and $w_i(x) = \sqrt{\frac{2}{\ell}}\ \sin(\frac{i\Pi x}{\ell})$. For the investigation of $u(t,x,\omega)$ the function $p_1(x)$ must not fulfil a condition but for the investigation of $u_{xxt}(t,x,\omega)$ we must assume that $p_1(x)$ fulfils the conditions $p_1 \in C^4$, $p_1(0) = p_1(\ell) = p_1''(0) = p_1''(\ell) = 0$.

We get very simple results for $p_1(x) = M\ \sin\ (\frac{m\Pi x}{\ell})$. It is

$$<\tilde{\zeta}_{o\ell}(t_1,\overset{1}{x})\tilde{\zeta}_{o\ell}(t_2,\overset{2}{x})> = \frac{aM^2\ell}{2}\ \frac{e^{-\beta s}}{4\mu_m}[\frac{1}{\beta}\cos\beta_m s+\frac{1}{\beta}\ \sin\beta_m s]D^{\ell}_{\overset{1}{x}}w_m(\overset{1}{x})D^{\ell}_{\overset{2}{x}}w_m(\overset{2}{x})$$

and

$$<\tilde{\zeta}_{1\ell}(t_1,\overset{1}{x})\tilde{\zeta}_{1\ell}(t_2,x)> = \frac{aM^2\ell}{2}\ \frac{e^{-\beta s}}{2}[\frac{1}{\beta}\cos\beta_m s-\frac{1}{\beta_m}\ \sin\beta_m s]D^{\ell}_{\overset{1}{x}}w_m(\overset{1}{x})D^{\ell}_{\overset{2}{x}}w_m(\overset{2}{x}).$$

The spectral density can be calculated by

$$S_{o\ell o\ell}(\alpha;\overset{1}{x},\overset{2}{x}) = \frac{M^2 a\ell}{4\Pi((\mu_m-\alpha^2)^2+4\beta^2\alpha^2)}\ D^{\ell}_{\overset{1}{x}}w_m(\overset{1}{x})D^{\ell}_{\overset{2}{x}}w_m(\overset{2}{x}).$$

In the case $p_1(x) = \delta(x-\overset{o}{x})$ we also obtain sensible results for u. This function p_1 describes an excitation which acts on $\overset{o}{x}$. The approximate correlation function of $u(t,x,\omega)$ is

$$<\tilde{\zeta}_{oo}(t_1,\overset{1}{x})\tilde{\zeta}_{oo}(t_2,\overset{2}{x})> = \sum_{i,j=1}^{\infty} T^{oo}_{ijo}(t_2-t_1)w_i(\overset{1}{x})w_j(\overset{2}{x})w_i(\overset{o}{x})w_j(\overset{o}{x})$$

$(t_1 < t_2)$ and its spectral density

$$S_{oooo}(\alpha;x,x) = 2a\ \sum_{i,j=1}^{\infty} W_{ij}(\alpha)w_i(x)w_j(x)w_i(\overset{o}{x})w_j(\overset{o}{x}).$$

Similarly, we can deal with many other concrete vibration problems which can be one-dimensional and also multidimensional relative to x. We can obtain a good summary of the random vibrations without many calculations if suitable special excitations are selected.

References

[1] Bogdanoff, J.L., Goldberg, J.E.: On the Bernoulli-Euler beam theory with random excitation. J.Aero/Space Sci.5,27 (1960), 371-376.

[2] Boyce, W.E.: Stochastic nonhomogeneous Sturm-Liouville problem. J.Franklin Inst.282. (1966), 206-215.

[3] Eringen, A.C.: Response of beams and plates to random loads. J. of Appl.Mech. 24, 1 (1957), 46-52.

[4] Purkert, W., vom Scheidt, J.: Limit theorems for solutions of stochastic differential equation problems. Intern.J.Math.and Math.Sci., Vol. 3, 1 (1980), 113-149.

[5] Purkert, W.. vom Scheidt, J.: Ein Grenzverteilungssatz für stochastische Eigenwertprobleme. ZAMM 59 (1979), 611-623.

[6] Uhlenbeck, G.E., Ornstein, L.S.: On the theory of the Brownian motion. Physical review, Vol. 36 (1930), 823-841.

Ingenieurhochschule

DDR 9500 Zwickau

Dr.Friedrichs-Ring 2a

German Democratic Republic

STOCHASTIC INTEGRATION WITH RESPECT TO MULTIPARAMETER GAUSSIAN PROCESSES

J.M.Stoyanov
Institute of Mathematics, Academy of Sciences

O.B.Enchev
Department of Mathematics, VIMMESS Institute
Rousse, Bulgaria

1. INTRODUCTION

This paper deals with stochastic integrals of the type $J^X(f)(\omega) = \int_T f(\vec{t})dX(\vec{t},\omega)$, where T is a set in the n- dimensional Euclidean space R^n, $f = (f(\vec{t}),\vec{t}\in T)$ is a real valued function and $X = (X(\vec{t}),t\in T)$ is a multiparameter random process.

The martingale theory of the stochastic integrals, i.e. the case when X is a semimartingale and f is random and X - adapted, will not be discussed.
Our aim here is to give a construction of the stochastic integral $J^X(f)$ for multiparameter f and X , and without using martingale type conditions. Some results in this direction could be found in [1, Ch.X], [6], [7] , [8], [9], [1o]. We shall present sufficient conditions for the existence of $J^X(f)$ using only some properties of the covariance function of the process X , plus an integrability of the function f .
Let us note that the standard approach for defining stochastic integrals is connected with considering Riemann-Stieltjes sums and passage to the limit in some sense. Here we shall use another approach based on the so-called convariance measures generated by the covariance function of the Gaussian process X . Namely, the stochastic integral $J^X(f)$ will be defined as the unique generalized Gaussian random process indexed by the elements of a suitable Hilbert space. Conditions for this are given in Theorem 1. Theorem 2 contains a formula for an effective calculation of the covariance of two integrals $J^X(f)$ and $J^X(g)$. Other propositions, as well as concrete examples are indicated.

The results of this paper are continuation of our previous studies [9], [1o] . In particular, the idea to use the covariance measures was presented by one of us /O.E./ at the 9th Spring Conference of the Bulgarian Mathematical Society, Sunny Beach, April 198o, see [11] . Recently, the same idea was applied successfully for studying stochastic integrals $J^X(f)$ for random functions f belonging to some Gaussian space generated by the Gaussian process X , see [12] .

2. SOME NOTATIONS. Suppose that all random variables and processes considered in this paper are defined on some complete probability space $(\Omega,\mathcal{F},\mathbb{P})$. For a convenience the expectation of the variable ξ will be denoted by $\langle\xi\rangle$.

If $\vec{a} = (a_1,\ldots,a_n)$ and $b = (b_1,\ldots,b_n)$ are two vectors in R^n, the symbol $\vec{a}<\vec{b}$ will mean that $a_1<b_1,\ldots,a_n<b_n$ (by analogy: $\vec{a}\le\vec{b}$ means $a_1\le b_1,\ldots, a_n\le b_n$). Hence we can define the set $[\vec{a},\vec{b}]$ in R^n as follows: $[\vec{a},\vec{b}] = \{\vec{t}\in R^n:\vec{a}\le\vec{t}\le\vec{b}\}$. Introduce the unit cell $T = [\vec{0},\vec{1}]$, where $\vec{0} = (0,0,\ldots,0)\in R^n$, $\vec{1} = (1,1,\ldots,1)\in R^n$.

In the sequel, $\mathcal{B}_T$ denotes the Borel σ-algebra of T, $\mathcal{C}(T)$ - the space of all real continuous functions on T, λ_m - the Lebesgue measure in R^m, $L_2(\Omega)$ - the space of all random variables with finite second moment.

3. EXISTENCE OF THE STOCHASTIC INTEGRAL. Let π be some subdivision of the cell $T = [\vec{0},\vec{1}]$ into the parallelepipeds $T_1^\pi, T_2^\pi,\ldots, T_{n(\pi)}^\pi$, where $n(\pi)$ is their number, T_k^π are closed, T_i^π and T_j, $i\neq j$, have not any common interior points and $|\pi|$ is the diameter of π. Let $\vec{\alpha}_{\pi,k}$ be an arbitrary point in T_k^π and $\mathcal{A}_\pi = \{\vec{\alpha}_{\pi,k}, k = 1,\ldots,n(\pi)\}$.
Now for any cell $I = [\vec{a},\vec{b}]\subset T$ and $k = 1,\ldots,n$, introduce two notations:

$$\boxed{k}_I X(\vec{t}) = X(t_1,\ldots,t_{k-1}, b_k, t_{k+1},\ldots,t_n)-$$

$$- X(t_1,\ldots,t_{k-1}, a_k, t_{k+1},\ldots,tn),$$

$$\Delta_I X = \boxed{1}_I \boxed{2}_I \ldots \boxed{n}_I X(\vec{t}).$$

The quantity

$$S_\pi = \sum_{k=1}^{n(\pi)} f(\vec{\alpha}_{\pi,k})\, \Delta_{T_k^\pi} X$$

is said to be a Riemann-Stieltjes sum for f and X in the multiparameter case. Notice that if $n=1$, then $S_\pi = \sum_{k=1}^{n(\pi)} f(\alpha_{\pi,k})[X(a_{\pi,k})-X(a_{\pi,k-1})]$,

where $\pi = \{a_{\pi,k}, k = 0,1,\ldots,n(\pi)\}$, $\alpha_{\pi,k}\in[a_{\pi,k-1}, a_{\pi,k}]$.

Definition 1. The mean square limit of S_π when $|\pi|\to 0$, if exists and does not depend on π and $\mathcal{A}_\pi$, is called a stochastic integral of the function f with respect to the random process X.
Notations: $\int_T f(\vec{t})dX(\vec{t},\omega)$, $J^X(f)$. ∎

For example, the papers [7], [9] contain concrete conditions under which the stochastic integral $J^X(f)$ exists for f and X with one-dimensional time parameter t.

Suppose X is a Gaussian process with zero mean, $\langle X(\vec{t})\rangle = 0$, and covariance function $K(\vec{s},\vec{t}) = \langle X(\vec{s})X(\vec{t})\rangle$, $\vec{s},\vec{t}\in T$.

The definition 1 implies, see [1], that $J^X(f)$ exists if and only if the integral $\int_{T\times T} f(\vec{s})f(\vec{t})d_{\vec{s}}d_{\vec{t}} K(\vec{s},\vec{t})$ exists. Also, if for some f and g the integrals $J^X(f)$ and $J^X(g)$ exist, then the integral $I_1 = \int_{T\times T} f(\vec{s})g(\vec{t})\, d_{\vec{s}}d_{\vec{t}} K(\vec{s},\vec{t})$ exists and

$$\langle J^X(f)\; J^X(g)\rangle = \int_{T\times T} f(\vec{s})g(\vec{t})d_{\vec{s}}d_{\vec{t}} K(\vec{s},\vec{t}).$$

Consider the vectors $\vec{a}$, $\vec{b}$, $\vec{a}'$, $\vec{b}' \in T$. Let the symbol $\{\vec{a}, \vec{b}\}$ denote a 2n-dimensional vector with first n components $\vec{a}$ and last n components $\vec{b}$. For $\{\vec{a}, \vec{b}\} \leq \{\vec{a}', \vec{b}'\}$ introduce in R^{2n} the cell $C = [\{\vec{a}, \vec{b}\}, \{\vec{a}', \vec{b}'\}]$ which is a subset of the 2n-dimensional unit cell $T\times T = [\{\vec{0},\vec{0}\}, \{\vec{1}, \vec{1}\}] = [(\underbrace{0,0,\ldots,0}_{2n}), (\underbrace{1,1,\ldots,1}_{2n})]$.

In the sequel we shall use the following notations:

$$\boxed{i}\,{}^{\vec{s}}_{C}\, K(\vec{s},\vec{t}) = K(s_1,\ldots,s_{i-1}, a_i', s_{i+1},\ldots,s_n, \vec{t}) -$$

$$- K(s_1,\ldots,s_{i-1},a_i, s_{i+1},\ldots,s_n, \vec{t}),$$

$$\boxed{i}\,{}^{\vec{t}}_{C}\, K(\vec{s},\vec{t}) = K(\vec{s},t_1,\ldots,t_{i-1}, b_i', t_{i+1},\ldots,t_n) -$$

$$- K(\vec{s},t_1,\ldots,t_{i-1}, b_i, t_{i+1},\ldots,t_n),$$

$$\Delta_C K = \boxed{1}\,{}^{\vec{s}}_{C}\,\boxed{1}\,{}^{\vec{t}}_{C} \ldots \boxed{n}\,{}^{\vec{s}}_{C}\,\boxed{n}\,{}^{\vec{t}}_{C}\, K(\vec{s}, \vec{t}).$$

One of the basic assumptions is the following $K(\vec{s},\vec{t})$, $\vec{s},\vec{t}\in T$, is a continuous function and has a bounded variation. Denote by $\mathcal{K}$ the class of all random processes, the covariance functions of which satisfy these two conditions.

Now let $X\in\mathcal{K}$ and K be its covariance function. Then [5] there exists a sign measure μ_k in the measurable space $(T\times T, \mathcal{B}_T\times\mathcal{B}_T)$ such that for any set of the type $(\vec{c}_1,\vec{c}_2)\subset T\times T$ we have

$$\mathcal{M}_K([\vec{c}_1,\vec{c}_2)) = \Delta_{[\vec{c}_1,\vec{c}_2)}K\ , \quad |\mathcal{M}_K|([\vec{c}_1,\vec{c}_2)) = \underset{[\vec{c}_1,\vec{c}_2)}{\mathrm{Var}}\ K.$$

This measure $\mathcal{M}_K$ will be called a <u>covariance measure</u> corresponding to the process X (also to the covariance function K).

Define a new measure $\tilde{\mathcal{M}}_K$ in $(T,\mathcal{B}_T)$ as follows:

$$\tilde{\mathcal{M}}_K(M) = \mathcal{M}_K(M\times T),\ M\in\mathcal{B}_T.$$

Let $L_2(T,\mathcal{B}_T,|\tilde{\mathcal{M}}_K|)$ be the Hilbert space of all functions defined on T, measurable about $\mathcal{B}_T$ and square integrable with respect to the measure $|\tilde{\mathcal{M}}_K|$. Denote this space by H. Evidently, for arbitrary $f,g\in H$, the following integral I_2 exists:

$$I_2 = \int_{T\times T} f(\vec{s})g(\vec{t})\mathcal{M}_K(d(\vec{s},\vec{t})).$$

Take now two functions $f,g\in\mathcal{C}(T)$. Then the integrals I_1 and I_2 exist and $I_1 = I_2$. Thus we come to the following statement.

PROPOSITION 1. <u>Let</u> $X\in\mathcal{K}$, $f,g\in\mathcal{C}(T)$. <u>Then the stochastic integrals</u> $J^X(f)$ <u>and</u> $J^X(g)$ <u>exist</u>. <u>Moreover, the Lebesgue integral</u> I_2 <u>also exists and</u>

$$\langle J^X(f)J^X(g)\rangle = \int_{T\times T} f(\vec{s})g(\vec{t})\mathcal{M}_K(d(\vec{s},\vec{t})).$$

<u>In particular</u>

(1) $$||J^X(f)||^2_{L_2(\Omega)} = \int_{T\times T} f(\vec{s})f(\vec{t})\mathcal{M}_K(d(\vec{s},\vec{t})).\ \boxtimes$$

In our further reasonings we shall use essentially the notation

$$B(f,g) = \int_{T\times T} f(\vec{s})g(\vec{t})\mathcal{M}_K(d(\vec{s},\vec{t})).$$

The equality (1) shows that $B(f,f)\geq 0$ for each $f\in\mathcal{C}(T)$. It is easy to see that the relation $B(f,f)\geq 0$ remains true for all $f\in H$. Hence $B(\cdot,\cdot)$ is a symmetric bilinear form, $B:H\times H\to R^1$, which is non-negative definite in the following sense: $B(f,f)\geq 0$, $f\in H$. Moreover, $|B(f,g)|\leq||f||_H\cdot||g||_H$. Now we can apply a corollary of the Riesz theorem [3] and to conclude that there exists a unique bounded linear opeator A: $H\to H$ such that $B(f,g) = (Af,g)_H$, $f,g\in H$.

Consider the set $A(H)\subset H$. It is pre-Hilbert space with an inner product $[\tilde{f},\tilde{g}]\equiv B(f,g) = B(g,f) = (Af,g)_H = (f,Ag)_H$, where $\tilde{f},\tilde{g}\in A(H)$,

$f\in A^{-1}\tilde{f}$, $g\in A^{-1}\tilde{g}$. Let $\mathcal{H}$ be the completion of A(H). It is not difficult to see that if the set H_1 is dense in H with respect to $||\cdot||_H$, then the set $A(H_1)$ is dense in A(H) with respect to the norm $||\cdot||_{\mathcal{H}}$ in $\mathcal{H}$. In particular, $A(\mathcal{E}(T))$ is dense in A(H) with respect to $||\cdot||_{\mathcal{H}}$.

Let us recall, see [4], that a generalized random process indexed by the elements of some real vector space V, is a mapping φ from V into the space of the random variable in $(\Omega,\mathcal{F},\mathbb{P})$ with the properties: for any v, v_1, $v_2\in V$ and $\alpha\in R^1$, $\mathbb{P}$-almost surely

$$\varphi(v_1 + v_2) = \varphi(v_1) + \varphi(v_2),\ \varphi(\alpha v) = \alpha\varphi(v).$$

Now, let be the Hilbert space defined above, and φ be the unique (up to isomorphism of the original probability space) generalized Gaussian random process indexed by the elements of $\mathcal{H}$, i.e. φ is a generalized random process and $\{\varphi(v),\ v\in\mathcal{H}\}$ is a complete system of Gaussian random variables with $\langle\varphi(u)\rangle = 0$, $u\in\mathcal{H}$, $\langle\varphi(u)\varphi(v)\rangle = (u,v)_{\mathcal{H}}$, $u,v\in\mathcal{H}$.

Definition 2. Let $X\in\mathcal{K}$, $f\in H$, A be the linear operator introduced above and φ be the unique generalized Gaussian process indexed by $\mathcal{H}$. Then the random variable $\varphi(Af)$ will be called stochastic integral of f with respect to X. ■

We have given two definitions of, maybe, one and the same object - the stochastic integral of f about X. The questions of the correctness and coicidence are quite natural. Using the linearity property of $J^X(f)$ (in both definitions) and the uniqueness of the generalized Gaussian process indexed by $\mathcal{H}$, we get the following statement.

PROPOSITION 2. The stochastic integral of the function f with respect to the Gaussian process X obtained by using Riemann-Stieltjes sums (Definition 1), if exists, coincides with that obtained as a generalized Gaussian process (Definition 2). ■

Let us note that Proposition 1 guarantees the existence of $J^X(f)$ only for functions $f\in\mathcal{E}(T)$. At the same time Definition 2 allows us to get $J^X(f)$ for an arbitrary $f\in H$. From this fact and Proposition 2 we come to a statement which is an extension of one of the basic results in [7].

THEOREM 1. Let the Gaussian process $X \in \mathcal{K}$ and f be an arbitrary function from H. Then the stochastic integral $J^X(f)$ is well defined. Moreover, for any $f, g \in H$, $\alpha, \beta \in R^1$, the following properties hold:

(i) $\langle J^X(f) \rangle = 0$;

(ii) $J^X(\alpha f + \beta g) = J^X(f) + \beta J^X(g)$;

(iii) $\langle J^X(f)\, J^X(g) \rangle = \int_{T\times T} f(\vec{s})g(\vec{t})\mathcal{M}_K(d(\vec{s},\vec{t}))$;

(iv) $||J^X(f)||^2_{L_2(\Omega)} = \int_{T\times T} f(\vec{s})f(\vec{t})\mathcal{M}_K(d(\vec{s},\vec{t}))$;

(v) the random variable $J^X(f)$ is distributed normally $\mathcal{N}(0, ||J^X(f)||^2)$;

(vi) $\{J^X(f),\ f \in H\}$ is a Gaussian system of random variables. ⊠

Proof. If we take into account Definition 2, we see that only (iii) must be verified. We have

$$\langle J^X(f)J^X(g) \rangle = \langle \varphi(Af)\varphi(Ag) \rangle = [Af, Ag]$$

$$= B(f,g) = \int_{T\times T} f(\vec{s})g(\vec{t})\mathcal{M}_K(d(\vec{s},\vec{t})). \quad ⊠$$

4. RULE FOR CALCULATING THE COVARIANCE OF TWO INTEGRALS. Theorem 1 establishes the existence of the stochastic integral $J^X(f)$ as an element of the space $L_2(\Omega)$. The relation (iii) shows how the covariance of two such integrals can be calculated. It is seen that for finding $\langle J^X(f)J^X(g) \rangle$ we must know something more about the structure of the covariance measure $\mathcal{M}_K$. Now we shall consider one concrete situation.

Assume that the covariance function K of the process $X \in \mathcal{K}$ satisfies the following conditions: 1°. All partial derivatives of $K(\vec{s},\vec{t})$ of order $\le 2n$ exist and are bounded in the domain $D = \{(\vec{s},\vec{t}) \in R^{2n} : \vec{s},\vec{t} \in T, s_1 \ne t_1, \ldots, s_n \ne t_n\}$; 2°. The derivative $\partial^{2n}K/\partial s_1 \partial t_1 \ldots \partial s_n \partial t_n$ is continuous in D, λ_{2n}-almost surely.

Let $N = \{1,2,\ldots,n\}$, $\Delta = \{i_1,\ldots,i_k\} \subset N$ and $|\Delta|$ be the number of the the elements of Δ, here $|\Delta| = K$, $K \le n$. Let

$$(T\times T)_{\Delta} = \{(\vec{s},\vec{t}) : \vec{s},\vec{t} \in T,\ s_i = t_i \text{ if } i \in \Delta, \text{ and } s_i \ne t_i \text{ if } i \notin \Delta\}.$$

Now we need the following notations:

$$\frac{\partial K(\vec{s},\vec{t})}{\delta(s_i=t_i)} = \frac{\partial K(\vec{s},\vec{t})}{\delta^-(s_i=t_i)} - \frac{\partial K(\vec{s},\vec{t})}{\delta^+(s_i=t_i)},$$

where

$$\frac{\partial K(\vec{s},\vec{t})}{\delta^-(s_i=t_i)} = \lim_{s_i \uparrow t_i} (s_i-t_i)^{-1}[K(s_1,\ldots,s_{i-1}, s_i, s_{i+1},\ldots s_n,\vec{t}) - K(s_1,\ldots,s_{i-1}, t_i, s_{i+1},\ldots s_n,\vec{t})],$$

$\frac{K(\vec{s},\vec{t})}{\delta^+(s_i=t_i)}$ is the same limit but for $s_i \downarrow t_i$.

It is easy to see that $\frac{\partial K}{\delta(s_i=t_i)}$ is the difference between the left and the right partial derivatives of K in its i-th argument.

For $\Delta = \{i_1,\ldots,i_k\}$ define the function $\delta\Delta K(\vec{s},\vec{t})$ as follows:

$$\delta^\Delta K = \frac{\partial^{2n-|\Delta|}}{\partial s_1 \partial t_1 \ldots \partial s_{i_1-1} \partial t_{i_1-1} \delta(s_{i_1}=t_{i_1}) \partial s_{i_1+1} \ldots \partial t_{i_k-1} \delta(s_{i_k}=t_{i_k}) \partial s_{i_k+1} \partial t_{i_k+1} \ldots \partial t_n}.$$

PROPOSITION 3. <u>The function $\delta^\Delta K$ is well defined in the set</u> $(T\times T)$. <u>More exactly, if $\Delta \neq \emptyset$, $\delta^\Delta K$ is bounded and continuous in</u> $(T\times T)_\Delta$, <u>and if $\Delta = \emptyset$, we have</u>

$$\delta^\Delta K(\vec{s},\vec{t}) = \frac{\partial^{2n} K(\vec{s},\vec{t})}{\partial s_1 \partial t_1 \ldots \partial s_n \partial t_n}$$

<u>which is λ_{2n} - a.s. continuous in</u> $D \equiv (T\times T)_\emptyset$. ∎

PROPOSITION 4. <u>The restriction of the measure $\mathcal{M}_K$ on $(T\times T)_\Delta$ is absolutely continuous with respect to the Lebesgue measure</u> $\lambda_{2n-|\Delta|}$. If $\frac{d\mathcal{M}_K}{d\lambda_{2n-|\Delta|}}$ <u>is the corresponding Radon-Nikodym derivative, then</u>

$$\frac{d\mathcal{M}_K}{d\lambda_{2n-|\Delta|}} = \delta^\Delta K. \quad ∎$$

The proofs of these propositions use the properties of the function K and are based on some statements from [2] and [7].

Thus we get the following result.

THEOREM 2. Let $X \in \mathcal{K}$ and its covariance function K satisfy the conditions 1°, 2° given above, and f, g be arbitrary functions from the space H. Then the stochastic integrals $J^X(f)$ and $J^X(g)$ exist and their covariance is expressed as follows:

$$<J^X(f)J^X(g)> = \sum_{\Delta:\Delta\subset N} \int_{(T\times T)_\Delta} f(\vec{s})g(\vec{t})\delta^\Delta K(\vec{s},\vec{t})\ \lambda_{2n-|\Delta|}(d(\vec{s},\vec{t})). \boxtimes$$

5. ILLUSTRATIONS. Now Theorem 2 will be illustrated by examples.

Example 1. Let n =1, T = [0,1]. Then

$$(2)\quad <J^X(f)J^X(g)> = \int_{[0,1]} f(t)g(t)\gamma(t)\lambda_1(dt)+ \int_{[0,1]^2} f(s)g(t)\frac{\partial^2 K}{\partial t\partial s}(s,t)\lambda_2(d(s,t)),$$

where $\gamma(t) = K'_\ell(t,t) - K'_r(t,t)$, K'_ℓ and K'_r are respectively the left and the right partial derivatives of K in its first argument. Actually, this is one of the main results in [7]. As we showed in [9] a similar formula to (2) is valid for random f, g not depending on X, and X not necessary Gaussian. In this case it is enough to replace fg by <fg>. □

Example 2. Let n = 2, T = [0,1]. Then

$$<J^X(f)J^X(g)> = \int_{[0,1]^4} f(s_1,s_2)g(t_1,t_2)\frac{\partial^4 K(s_1,s_2,t_1,t_2)}{\partial s_1\partial t_1\partial s_2\partial t_2}\ \lambda_4(d(s_1,s_2,t_1,t_2))$$

$$+ \int_{[0,1]^3} f(s_1,s_2)g(s_1,t_2)\frac{\partial^3 K(s_1,s_2,s_1,t_2)}{\delta(t_1=s_1)\partial s_2\partial t_2}\ \lambda_3(d(s_1,s_2,t_2))$$

$$+ \int_{[0,1]^3} f(s_1,s_2)g(t_1,s_2)\frac{\partial^3 K(s_1,s_2,t_1,s_2)}{\partial s_1\partial t_1\delta(t_2=s_2)}\ \lambda_3(d(s_1,s_2,t_1))$$

$$+ \int_{[0,1]^2} f(s_1,s_2)g(s_1,s_2)\frac{\partial^2 K(s_1,s_2,s_1,s_2)}{\delta(t_1=s_1)\delta(t_2=s_2)}\ \lambda_2(d(s_1,s_2)). \blacksquare$$

Example 3. Let n = 2, T = [0,1] and X be the process of Levy-Chentsov, its covariance function is $K(s_1,s_2,t_1,t_2) = \min[s_1,t_1]. \min[s_2,t_2]$.

Then

$$<J^X(f)J^X(g)> = \int_{[0,1]^2} f(s_1,s_2)g(s_1,s_2)\lambda_2(d(s_1,s_2)).$$

More generally, for the n-parameter Levy-Chentsov process, its covariance function is $K(\vec{s},\vec{t}) = \prod_{i=1}^{n} \min[s_i,t_i]$, we have

$$<J^X(f)J^X(g)> = \int_{[0,1]^n} f(\vec{s})g(\vec{s})\lambda_n(d(\vec{s})). \boxtimes$$

Jordan M.Stoyanov

Institute of Mathematics,
Academy of Sciences
1090 Sofia, P.O.Box 373,
Bulgaria

Ognian B.Enchev

Department of Mathematics,
VIMMESS Institute
7004 Rousse, Bulgaria

R E F E R E N C E S

[1] M.Loeve: "Probability Theory. 3rd ed." Van Nostrand, Princeton, 1963

[2] W.Rudin: "Principles of Mathematical Analysis". McGrav-Hill, New York, 1964

[3] M.Reed - B.Simon: "Methods of Modern Mathematical Physics. Volume 1." Academic Press, New York, 1972

[4] B.Simon: "The $P(\rho)_2$ Euclidian /Quantum/ Field Theory." Princeton Univ. Press, Princeton, 1974.

[5] G.Ciucu - C.Tudor: "Probabilitati si Procese Stocastice. Volume 1" Ed.Academiei, Bucuresti, 1978

[6] Yu.Daleckii - S.Paramonova: Stochastic Integrals with Respect to Normal Distributed Additive Set Functions. Doklady Acad.Nauk SSSR 2o8 , 512-515 /1973/.

[7] J.Yeh: Stochastic Integral of L_2-function with Respect to Gaussian Process. Tohoku Math.J.27, 175-186 /1975/.

[8] S.Huang - S.Cambanis: Stochastic and Multiple Wiener Integrals for Gaussian Processes. Ann.Probability 6, 585-614 /1978/.

[9] J.Stoyanov - O.Enchev: Stochastic Integrals for Gaussian Random Processes. C.R.Acad.Bulg.Sci. 32, 1467-147o /1979/.

[1o] O.Enchev - J.Stoyanov: Stochastic Integrals for Gaussian Random Functions. Stochastics 3, 277-289 /198o/.

[11] O.Enchev: Stochastic Integrals with Respect to Multiparameter Gaussian Random Processes. In the book: "Mathematics and Mathematical Education. Proc.9th Spring Conference of the Bulg. Math.Society, Sunny Beach, April, 198o". pp.128-132.

[12] O.Enchev: Integration with Respect to Wick Powers of Gaussian Random Measures. Differential Formula. C.R.C.Acad.Bulg.Sci. Vol.34 /1981/ N°.6.

ON L^2 AND NON-L^2 MULTIPLE STOCHASTIC INTEGRATION

D. Surgailis
Institute of Mathematics and Cybernetics, Vilnius

0. Introduction. In their pioniering works K.Ito [4], [5] and Wiener [17] developed the theory of multiple stochastic integrals (m.s.i.)

$$I^{(n)}(f) = \int_{T^n} f(t_1,\ldots,t_n)Z(dt_1)\ldots Z(dt_n) \tag{0.1}$$

where $f \in L^2(T^n)$ is a deterministic function on $T^n = T\times\ldots\times T$ (n times) (usually $T = \mathbb{R}^d$, $d\geq 1$), and $Z = Z(dt)$ is Gaussian or Poisson noise in T. (By noise we mean a finitely additive random set function with independent values on nonintersecting sets). A fundamental feature of the classical Ito-Wiener m.s.i. is their orthogonality and completeness: any random variable (r.v.) $\xi \in L^2(\Omega)$ can be uniquely expanded in orthogonal series of m.s.i., convergent in $L^2(\Omega)$.

There are however other motives to discuss m.s.i. apart from orthogonal expansions. One of them is the possibility to construct by them probabilistic objects with certain desirable properties as in the case of self-similar random fields [2], [14], [15].

Therefore it seems natural to consider also

(a) m.s.i. with respect to more general L^2-noises (i.e. noises with finite variance) that Gaussian or Poisson, which are $L^2(\Omega)$-r.v., or

(b) m.s.i. with respect to L^2-noises which are not $L^2(\Omega)$-r.v. or m.s.i. with respect to non-L^2-noises (among which stable noises are most important).

We shall refer to (a) as L^2-m.s.i. and to (b) as non-L^2 m.s.i. Generalization of the classical definition of m.s.i. in case (a) is straightforward, the only new problem arising is the completeness of the system of such m.s.i. Moreover, m.s.i. of type (a) can be reduced to classical m.s.i. with respect to Gaussian and Poisson noise in $T\times R$ [5] as Z itself can be represented in such a way. Nevertherless we regard the question of completeness of m.s.i. in case (a) worth discussion, especially as a clear answer to it in terms of characteristics of Z can be obtained (see below).

Let us briefly review the content of the remaining sections. In Sect. 1 we define in the usual way (via integral sums with discarded diagonals) m.s.i.

$$(0.2)\quad I^{(n)}(f) = \int_{T^n} \overline{\sum_{i_1,\dots,i_n=\overline{1,m}}} f_{i_1,\dots,i_n}(t_1,\dots,t_n) Z_{i_1}(dt)\dots Z_{i_n}(dt_n)$$

with respect to $\mathbb{R}^m$-valued L^2-noise $Z = (Z_1,\dots,Z_m)$ in T with characteristic function (ch.f.)

$$(0.3)\qquad E[\exp\{i(a,Z(A))\}] = \exp\{ \int_A w(t,a)\mu(dt)\}, \quad a\in \mathbb{R}^m,$$

where

$$(0.4)\qquad w(t,a) = -1/2\ (\sigma(t)a,a) + \int_{\mathbb{R}^m\setminus\{0\}} (e^{i(a,u)}-1-i(a,u))\pi(t,du)$$

under general conditions on the 'reference measure' $\mu(dt)$, 'diffusion matrix' $\sigma(t) = (\sigma_{ij}(t))_{i,j=\overline{1,m}}$ and the 'Lévy measure' $\pi(t,du)$. Integral (0.2) is well-defined for any (measurable) function $f = (f_{i_1,\dots,i_n}(t_1,\dots,t_n))_{i_1,\dots,i_n=\overline{1,m}}$, $t_1,\dots,t_n\in T$ such that

$$(0.5)\qquad ||f||_n = \Big(\int_{T^n} \overline{\sum_{\substack{i_1,\dots,i_n=\overline{1,m}\\ j_1,\dots,j_n=\overline{1,m}}}} f_{i_1,\dots,i_n}(t_1,\dots,t_n)\, \overline{f_{j_1,\dots,j_n}(t_1,\dots,t_n)}\, r_{i_1j_1}(t)\dots r_{i_nj_n}(t_n)d\mu^n\Big)^{1/2}<+\infty$$

where $r_{ij}(t) = E[Z_i(dt)Z_j(dt)]/\mu(dt)$, $i,j=1,\dots,m$ is the covariance (density) matrix of Z, denoted by $R(t)=(r_{ij}(t))_{i,j=\overline{1,m}}$. The basic properties of m.s.i. (0.2) will be discussed.

In Sect. 2 we investigate the completeness of m.s.i. system (0.2) in $L^2(\Omega) = L^2(\Omega,F,P)$, where F is the σ-algebra generated by Z. It turns out that this system is complete iff the equality

$$(0.6)\qquad 2\mathrm{Re}w(t,a) + (R^{-1}(t)\mathrm{grad}w(t,a),\mathrm{grad}w(t,a)) = 0,\ a\in\mathbb{R}^m$$

holds $d\mu$-a.e. in T (Th.2.4) (grad in (0.6) refers to differentiation with respect to $a\in\mathbb{R}^m$). In any case, the left hand side of (0.6) is negative (≤ 0). This inequality implies another interesting one:

$$(0.7)\qquad |\varphi(a)|^2 + (S^{-1}\mathrm{grad}\varphi(a),\mathrm{grad}\varphi(a)) \le 1,\ a\in\mathbb{R}^n .$$

Here $\varphi(a) = E[\exp\{i(a,\xi)\}]$ is ch.f. of arbitrary r.v.ξ with values in R^n with zero mean and covariance matrix S, det S $\neq$ 0. An elementary proof of (0.7), in case n=1 due to A.Grincevičius, is presented in Appendix. In the scalar case m=1 the system of m.s.i. (0.1) is complete in $L^2(\Omega)$ iff $d\mu$-a.e. in T either $\pi(t,\cdot)=0$ or $\sigma(t)=0$ and $\pi(t,du)$ is concentrated in one point (Th.2.6).

In Sect. 3 we discuss m.s.i. by scalar stable noises with ch.f.

$$(0.8)\qquad E[\exp\{iaZ(A)\}] = \exp\{-|a|^{\alpha}\mu(A)\},\quad a\in\mathbb{R},$$

where $1<\alpha<2$. A plausible hypothesis says that m.s.i. (0.1) with respect to such Z can be defined for any $f\in L^{\alpha}(T^n)$ and should belong to $L^{\beta}(\Omega)$ for all $\beta<\alpha$(eventually to weak Lebesgue space $L^{\alpha}_{w}(\Omega)$). However, we establish a weaker result. Let $f=f(t_1,\ldots,t_n)$ be a symmetric complex function on T^n such that

$$(0.9)\qquad \Big(\sum_{k=0}^{n}\int_{T^k}\Big(\int_{T^{n-k}}|f(t_1,\ldots,t_k,\ldots,t_n)|^{\alpha+\varepsilon}\mu(dt_{k+1})\ldots\mu(dt_n)\Big)^{\frac{\alpha-\varepsilon}{\alpha+\varepsilon}}\cdot\big(\mu(dt_1)\ldots\mu(dt_k)\big)^{\frac{1}{\alpha-\varepsilon}}<+\infty$$

for some $\varepsilon > 0$. Then (0.1) is well-defined and $E[|I^{(n)}(f)|^{\alpha-\varepsilon}]<+\infty$ (Th.3.2). We apply this result to define a new class of 'polynomial' self-similar processes subordinated to stable noise in R (Th.3.3).

L^2-m.s.i. with respect to Gaussian and Poisson noises and their analogs ('Wick polynomials') were the subject of numerous investigations (see e.g. [2], [3], [6], [7], [10], [13]). In connection with the non-L^2 case, the martingale approach to multiple stochastic integration (see eg. [9]) seems to be promising but we did not follow it. While this paper was in preparation, the author obtained a letter from prof. M.S.Taqqu saying that he and prof.R.Wolpert were working on non-L^2 multiple integration with respect to stable noises but no details were given.

Acknowledgements. The author is grateful to A.Astrauskas and R.Mikulevicius for fruitful discussions.

1. L^2-multiple stochastic integrals: definition and basic properties.

Let $T\subset R^d$ $(d\geq 1)$ be an open set, B(T) be its Borel subsets and $B_c(T)$ its relatively compact subsets. Let μ be a σ-finite measure on T such that $\mu(A)<+\infty$ for any $A\in B_c(T)$. Finally, let be given a $\mathbb{R}^m$-valued noise

$Z = (Z_1,\dots,Z_m)$ in T, i.e. a family of random vectors $Z(A)=(Z_1(A),\dots, Z_m(A))$, $A\in B_c(T)$ defined on a probability space (Ω,F,P) such that for any nonintersecting sets $A_1,\dots,A_n\in B_c(T)$, $n\geq 1$, $Z(A_1),\dots,Z(A_n)$ are independent and $Z(A_1)+\dots+Z(A_n) = Z(\bigcup_{k=1}^{n} A_k)$. We assume that ch.f. of $Z(A)$ is given by (0.3), (0.4), where $\sigma(t)=(\sigma_{ij}(t)): T \to \mathbb{R}^m\otimes\mathbb{R}^m$ is measurable [1] and positive definite $d\mu$-a.e. in T, and $\pi(t,\cdot)$ is a kernel on $T\times B(\mathbb{R}^m\times\{0\})$ such that

$$(1.1)\qquad \int_A (\operatorname{trace}\sigma(t) + \int_{\mathbb{R}^m\setminus\{0\}} |u|^2\,\pi(t,du))\mu(dt)<+\infty,\ A\in B_c(T)\cdot$$

We assume also that μ is diffuse: $\mu(\{t\}) = 0\ \forall t\in T$ and the basic σ-algebra F is generated by r.v. $Z(A)$, $A\in B_c(T)$. By (0.3), (0.4),

$$(1.2)\qquad E[Z_i(A)Z_j(A)] = \int_A r_{ij}(t)\mu(dt),\ A\in B_c(T),\ i,j=1,\dots,m,$$

where

$$(1.3)\qquad r_{ij}(t) = \sigma_{ij}(t) + \int_{\mathbb{R}^m\setminus\{0\}} u_i u_j\,\pi(t,du).$$

While some of the above conditions are not essential and can be weakened, the following one is more restrictive: (R) the covariance matrix $R(t) = (r_{ij}(t))$ is strictly positive definite $d\mu$-a.e. in T. The above assumptions including (R) are assumed to hold in Sections 1-2 without explicit reference to them.

Introduce the Hilbert space $L^2(T)$ of functions $f:T\to\mathbb{C}^m$ such that

$$(1.4)\qquad ||f|| = (\int_T (R(t)f(t),f(t))\mu(dt))^{1/2} < +\infty\ .$$

By (R), $L^2(T)$ is complete if we identify as usual functions which are equal $d\mu$-a.e. It can be shown that n-tuple tensor product $(\otimes L^2(T))^n$ can be identified with the Hilbert space $L^2(T^n)$ consisting of all functions $f:T^n\to(\otimes\mathbb{C}^m)^n$, $f=(f_{i_1,\dots,i_n})_{i_1,\dots,i_n=\overline{1,m}}$ with finite norm $||f||_n$ (0.5), while symmetric tensor product $(\hat{\otimes}L^2(T))^n$ - with the subspace $L_s^2(T^n)\subset L^2(T^n)$ consisting of symmetric functions: $f = \operatorname{sym}f$, where

$$(\operatorname{sym}f)_{i_1,\dots,i_n}(t_1,\dots,t_n) = \Sigma f_{i_{\pi(1)},\dots,i_{\pi(n)}}(t_{\pi(1)},\dots,t_{\pi(n)})/n!$$

1 We shall not mention measurability below when this will be clear from the context.

and the sum is taken over all permutations $\pi(1),\ldots,\pi(n)$ of $1,\ldots,n$. Finally, introduce Fock space $F(L^2(T))$ as the space of all sequences $f=(f_0,f_1,\ldots)$ such that $f_n \in L^2_s(T^n), n\geq 0$ and

$$(1.5) \qquad ||f||_F = \left(\sum_{n=0}^{\infty} ||f_n||_n^2 / n!\right)^{1/2} < +\infty$$

(we set $L^2(T^0)=\mathbb{C}$ and $||f||_0=|f|$). Denote also $(\cdot,\cdot)_n$, $(\cdot,\cdot)_F$ the scalar products in $L^2(T^n)$ and $F(L^2(T))$, respectively.

Theorem 1.1. For any $n\geq 1$ and any $f=(f_{i_1,\ldots,i_n}) \in L^2(T^n)$ there exists r.v. $I^{(n)}(f)$ called multiple stochastic integral (m.s.i.) of f with respect to Z, denoted also by (0.2), with the following properties:

(i1) $I^{(n)}(f) = I^{(n)}(\mathrm{sym} f) \in L^2(\Omega)$;

(i2) $E[I^{(n)}(f)] = 0$;

(i3) $E[I^{(n)}(f)\overline{I^{(k)}(g)}] = \delta_{nk}\, n!(\mathrm{sym} f, g)_n$ for any $k\geq 1$ and $g\in L^2(T^k)$, where δ_{nk} is Kronecker's symbol.

Proof. Let $(\Delta)_N$, $N=1,2,\ldots$ be a sequence of partitions of T by sets $\Delta\in B_c(T)$ such that $(\Delta)_N \subset (\Delta)_{N+1}$ and $\sup(\mathrm{diam}\Delta \,|\, \Delta\in(\Delta)_N) \to 0$ $(N\to\infty)$. A function $f=(f_{i_1,\ldots,i_n}) \in L^2(T^n)$ is said simple if (a) f is constant on subsets $D\subset T^n$ of the form $D=\Delta_1\times\ldots\times\Delta_n$, $\Delta_1,\ldots,\Delta_n\in(\Delta)_N$, for some $N\geq 1$ and vanishes on a finite number of such D's, and (b) f vanishes on 'diagonals': $f^{\Delta_1,\ldots,\Delta_n}_{i_1,\ldots,i_n} = 0$ if $\Delta_i = \Delta_j$ for $i\neq j$, $i,j=1,\ldots,n$, $i_1,\ldots,i_n=1,\ldots,m$, where $f^{\Delta_1,\ldots,\Delta_n}_{i_1,\ldots,i_n}$ is the value $f_{i_1,\ldots,i_n}$ taken on $\Delta_1\times\ldots\times\Delta_n$. For simple $f\in L^2(T^n)$ set

$$(1.6) \qquad I^{(n)}(f) = \sum_{\Delta_1,\ldots,\Delta_n\in(\Delta)_N} \; \sum_{i_1,\ldots,i_n=\overline{1,m}} f^{\Delta_1,\ldots,\Delta_n}_{i_1,\ldots,i_n} Z_{i_1}(\Delta_1)\ldots Z_{i_n}(\Delta_n).$$

The right hand side of (1.6) is well-defined for N sufficiently large and does not depend on N. It can be easily verified that (1.6) satisfies (i1)-(i3). For arbitrary $f\in L^2(T^n)$, set $I^{(n)}(f) = \mathrm{l.i.m.}\; I^{(n)}(f_k)$ $(k<\infty)$, where $(f_k)_{k\geq 1}$ is a sequence of simple functions convergent to f in $L^2(T^n)$. Due to (i3), such a limit exists and satisfies (i1)-(i3) as well. ∎

Corollary 1.2. Denote $L_I^2(\Omega)$ the Hilbert subspace of $L^2(\Omega)$ spanned by r.v. $I^{(n)}(f)$, $f \in L^2(T^n)$, $n \geq 0$ (we set $I^{(o)}(f)=f$, $f \in \mathbb{C}$). Then $L_I^2(\Omega)$ is unitarily isomorphic to $F(L^2(T))$, the unitary $I: F(L^2(T)) \to L_I^2(\Omega)$ being given by

$$(1.7) \qquad I(f) = \sum_{n=0}^{\infty} I^{(n)}(f_n)/n!, \quad f=(f_o, f_1, \ldots) \in F(L^2(T)).$$

Remark 1.3. Given a component $f_{j_1,\ldots,j_n}$ of $(\otimes \mathbb{C}^m)^n$-valued function $f=(f_{i_1,\ldots,i_n}) \in L^2(T^n)$, the integral $\int_{T^n} f_{j_1,\ldots,j_n}(t_1,\ldots,t_n) \cdot Z_{j_1}(dt_1)\ldots Z_{j_n}(dt_n)$ is not in general well-defined as the function $\tilde{f}=(\tilde{f}_{i_1,\ldots,i_n}): T^n \to (\otimes \mathbb{C}^m)^n$, $\tilde{f}_{i_1,\ldots,i_n} = f_{j_1,\ldots,j_n}$ if $(i_1,\ldots,i_n) = (j_1,\ldots,j_n)$, $= 0$ if otherwise might be not in $L^2(T^n)$. This is true, however, if

$$(1.8) \quad ||g|| \geq c \left(\int_T \sum_{i=1}^{n} r_{ii}(t)|g_i(t)|^2 \, d\mu\right)^{1/2}, \quad g = (g_1,\ldots,g_m) \in L^2(T),$$

as (1.8) implies

$$(1.9) \quad \begin{aligned} ||f||_n \geq c_n \Big(\int_{T^n} \sum_{i_1,\ldots,i_n=\overline{1,m}} & |f_{i_1,\ldots,i_n}(t_1,\ldots,t_n)|^2 \cdot \\ & \cdot r_{i_1 i_1}(t_1)\ldots r_{i_n i_n}(t_n) d\mu^n\Big)^{1/2}, \quad f \in L^2(T^n), \end{aligned}$$

where c, c_n are some constants.

2. Completeness of the system of m.s.i. In this Section we discuss the question when $L^2(\Omega) = L_I^2(\Omega)$. First we need some auxiliary lemmas.

Lemma 2.1. Let $\xi_1, \xi_2, \ldots$ be a sequence of real r.v., $F_\infty = \sigma(\xi_1, \xi_2, \ldots)$. Then the linear combinations $\sum_{k=1}^{n} c_k \exp\{i \sum_{j=1}^{n} \lambda_{kj} \xi_j\}$, $c_k \in \mathbb{C}$, $\lambda_{kj} \in \mathbb{R}$, $k,j = 1,\ldots,n$, $n \geq 1$ are dense in $L^2(\Omega, F_\infty, P)$.

Proof is easy and therefore omitted. ⊠

Denote $\mathrm{Re}L^2(T)$ the subspace of functions $f \in L^2(T)$ with real components $f_1,\ldots,f_m$. Lemma 2.1 implies easily

Lemma 2.2. $L^2(\Omega) = L_I^2(\Omega)$ iff for any $f \in \mathrm{Re}L^2(T)$, $\exp\{iI^{(1)}(f)\} \in L_I^2(\Omega)$.

Lemma 2.3. Let $f \in \mathrm{Re}L^2(T)$, $h \in L^2(T^n)$, $n \geq 1$, and let (1.8) hold. Then

$$E[e^{iI^{(1)}(f)}\overline{I^{(n)}(h)}] =$$

(2.2)

$$E[e^{iI^{(1)}(f)}]\int_{T^n} \overline{\sum_{i_1,\ldots,i_n=1,m}} \prod_{k=1}^{n} g(t_k,f)_{i_k}\, h_{i_1,\ldots,i_n}(t_1,\ldots,t_n)d\mu^n,$$

where

(2.3) $$g(t,f) = \int_{\mathbb{R}^m\setminus\{0\}} u(e^{i(f(t),u)}-1)\pi(t,du) + i\sigma(t)f(t).$$

Proof. By (1.8) and (1.9), (2.2) is well-defined and continuously dependent on f and h with respect to convergence in $L^2(T)$ and $L^2(T^n)$ respectively, hence it suffices to prove (2.2) for simple f and h. This follows directly from formulae (0.3),(0.4) of ch.f. of Z and definition (1.6). (See also [3], eq. (5) § 11 or [13], Prop. 2.2 for analogous formula in case of scalar Poisson noise Z). ∎

Theorem 2.4. Let (1.8) hold. Then $L^2_I(\Omega) = L^2(\Omega)$ if and only if equality (0.6) holds $d\mu$-a.e. in T.

Proof is based on Parseval's equality. Let us construct orthonormal (O.N.) basis in $L^2_I(\Omega)$ by means of isometry I (1.7). If $g^{(1)},\ldots, g^{(n)} \in L^2(T)$, then $g^{(1)}\otimes\ldots\otimes g^{(n)}$ is in $L^2(T^n)$ and

(2.4)
$$(g^{(1)}\otimes\ldots\otimes g^{(n)})_{i_1,\ldots,i_n}(t_1,\ldots,t_n) = g^{(1)}_{i_1}(t_1)\ldots g^{(n)}_{i_n}(t_n),$$
$$i_1,\ldots,i_n = 1,\ldots,m.$$

Let $k_1,k_2,\ldots$ be nonnegative integers such that $\Sigma k_i < +\infty$ (i.e. only finite number of them are not zero). Let $(f^{(j)})_{j\geq 1}$ be O.N. basis in $L^2(T)$. Introduce functions

(2.5) $$h^{(k_1,k_2,\ldots)} = (\otimes f^{(1)})^{k_1}\hat{\otimes}(\otimes f^{(2)})^{k_2}\hat{\otimes}\ldots \in L^2_s(T^n),$$

where $n = \Sigma k_i$, $(\otimes f)^0 = 1$, $f\otimes 1 = f$. Identify $f \in L^2_s(T^n)$ with element $(\underbrace{0,\ldots,0}_{n-1},f,0,\ldots) \in F(L^2(T))$. By definition,

(2.6) $$(h^{(k_1,k_2,\ldots)},h^{(k'_1,k'_2,\ldots)})_F = \delta_{nn'}\prod_{j=1}^{\infty}\delta_{k_jk'_j}\, k_j!/(n!)^2,$$

$n=\Sigma k_i$, $n'=\Sigma k'_i$. Therefore the system $\tilde{h}^{(k_1,k_2,\ldots)} = (n!/\sqrt{k_1!k_2!\ldots})\times$

$\times h^{(k_1,k_2,\ldots)}$, $k_1\geq 0, k_2\geq 0,\ldots$ is O.N. in $F(L^2(T))$ and actually O.N. basis. The corresponding O.N. basis in $L_I^2(\Omega)$ is given by

$$I(\tilde{h}^{(k_1,k_2,\ldots)}) = I^{(n)}(h^{(k_1,k_2,\ldots)})/\sqrt{k_1!\,k_2!\ldots},$$

(2.7)

$$n = \Sigma k_i,\ k_1,k_2,\ldots\geq 0.$$

By (2.2), (2.4) and (i1),

$$c_{k_1,k_2,\ldots}(f) \equiv E[e^{iI^{(1)}(f)}\overline{I(\tilde{h}^{(k_1,k_2,\ldots)})}]$$

(2.8)

$$= E[e^{iI^{(1)}(f)}]\prod_{j=1}^{\infty}(\int_T(g(t,f),f^{(j)}(t))\mu(dt))^{k_j}/\sqrt{k_j!},$$

hence

$$|c(f)|^2 = \sum_{k_1,k_2,\ldots\geq 0}|c_{k_1,k_2,\ldots}(f)|^2$$

(2.9)

$$= |E[e^{iI^{(1)}(f)}]|^2 \exp\{\int_T(R^{-1}(t)g(t,f),g(t,f))d\mu\}$$

as $(f^{(j)})_{j\geq 1}$ is O.N. basis in $L^2(T)$. Observe that $ig(t,a) = \mathrm{grad}w(t,a)$ and $|E[e^{iI^{(1)}(f)}]|^2 = \exp\{2\mathrm{Re}\int_T w(t,f)d\mu\}$. By Parseval's equality, $\exp\{iI^{(1)}(f)\}\in L_I^2(\Omega)$ is equivalent to $|c(f)|^2 = 1$. As $f\in \mathrm{Re}L^2(T)$ is arbitrary, Lemma 2.2 and (2.9) imply Theorem 2.4. ∎

By the same argument and Bessel's inequality the left hand side of (0.6) is negative (≤ 0):

(2.10) $\quad (R^{-1}(t)\mathrm{grad}w(t,a),\mathrm{grad}w(t,a)) + 2\mathrm{Re}w(t,a) \leq 0,\ a\in\mathbb{R}^m.$

In particular, let $w(a) = \int_{\mathbb{R}^{n+1}\setminus\{0\}}(e^{i(a,u)}-1-i(a,u))\pi(du)$, $a\in\mathbb{R}^{n+1}$, where $\Pi = m\times\delta_{\{1\}}$, m is a probability measure on $\mathbb{R}^n$ with zero mean $\int_{\mathbb{R}^n} x_i m(dx) = 0$, $i=1,\ldots,n$, and covariance matrix S, det S $\neq 0$, and $\delta_{\{1\}}$ is unite mass concentrated at $1\in\mathbb{R}$. Set $r_{ij} = \int_{\mathbb{R}^{n+1}} u_i u_j \pi(du)$, $R=(r_{ij})_{i,j=\overline{1,n+1}}$, $\tilde{a}=(a_1,\ldots,a_n,0)\in\mathbb{R}^{n+1}$, $a=(a_1,\ldots,a_n)\in\mathbb{R}^n$, $\varphi(a) = \int_{\mathbb{R}^n} e^{i(a,x)}m(dx)$, $a\in\mathbb{R}^n$ is ch.f. of m. Then $2\,\mathrm{Re}w(\tilde{a})=2\mathrm{Re}\varphi(a) - 2$,

$$(R^{-1}\text{grad}w(\tilde{a}),\text{grad}w(\tilde{a})) = (S^{-1}\text{grad}\varphi(a),\text{grad}\varphi(a)) + |\varphi(a)-1|^2 .$$

Together with (2.10) this implies.

Corollary 2.5. Ch.f. $\varphi(a)$, $a\in\mathbb{R}^n$ of arbitrary probability distribution in $\mathbb{R}^n$ with zero mean and nondegenerate covariance matrix S satisfies inequality (0.7).

In particular if n=1 we have

$$|\varphi(a)|^2 + |\varphi'(a)|^2/\sigma^2 \le 1, \ a\in\mathbb{R} \qquad (2.11)$$

for any ch.f. φ such that $\varphi'(0)=0$ and $-\varphi''(0)= \sigma^2 < +\infty$.

Inequalities (0.7), (2.11) seem to be new. For another proof of (2.11) see Appendix.

Coming back to Th. 2.4, eq.(0.6) is satisfied in the Gaussian case ($\pi(t,\cdot)=0$). If m=1, we have the following alternative

Theorem 2.6. Let m=1 in Th.2.4 (in such a case (1.8) is satisfied automatically). Then $L^2_I(\Omega) = L^2(\Omega)$ iff the following alternative holds $d\mu$-a.e. in T: either $\pi(t,\cdot)=0$ or $\sigma(t)=0$ and $\pi(t,\cdot)$ is concentrated in one point.

We omit the proof of this theorem which is based on Th.2.4 and the following fact: let $u_1,u_2\neq 0$ be real numbers such that $iu_j = c(e^{iu_j}-1)$, j=1,2 with some $c\in\mathbb{C}$, then $u_1=u_2$. Let us note that an analog statement of Th.2.6 is not correct if m>1. For example, let m=2, T=(0,1), $\mu(dt)=dt$, $\sigma(t)=0$ and $\pi(t,\cdot) = \delta_{\{1,1\}} + \delta_{\{-1,1\}}$. Then (0.6) holds and consequently $L^2_I(\Omega) = L^2(\Omega)$.

3. Multiple integrals with respect to stable noises. In this Section we discuss non-L^2 m.s.i. with respect to Poisson and stable noises.

Let T,μ be the same as above, and let be given (scalar) Poisson noise $p = p(A)$, $A\in B_c(T)$ in T with mean $E[p(A)] = \mu(A)$. Set $q(A) = p(A)-\mu(A)$. If $f:T^n \to \mathbb{C}$ is simple function, then n-tuple s.i. of f with respect to q is equal to

$$Q^{(n)}(f) = \sum_{\Delta_1,\dots\Delta_n} f^{\Delta_1,\dots,\Delta_n}(p(\Delta_1)-\mu(\Delta_1))\dots(p(\Delta_n)-\mu(\Delta_n)).$$

Therefore

$$E[|Q^{(n)}(f)|] \le 2^n \int_{T^n} |f(t_1,\dots,t_n)|d\mu^n = 2^n ||f||_{L^1(T^n)}. \tag{3.1}$$

On the other hand, by (i3),

$$E^{1/2}[|Q^{(n)}(f)|^2] \le \sqrt{n!}\,||f||_{L^2(T^n)}. \tag{3.2}$$

Applying the Riesz-Thorin interpolation [12] we have

Theorem 3.1. Let p be Poisson noise in T with mean μ, $q=p-\mu$, and let be given a sequence $(f_k)_{k\ge 1}$ of simple functions convergent to f in $L^p(T^n)$, $1\le p\le 2$, $n\ge 1$. Then the integral sums $Q^{(n)}(f_k)$ converge in $L^p(\Omega)$ as $k\to\infty$ to r.v. $Q^{(n)}(f) = \int_{T^n} f(t_1,\dots,t_n)q(dt_1)\dots q(dt_n)$ and

$$E^{1/p}[|Q^{(n)}(f)|^p] \le C||f||_{L^p(T^n)},\quad C = 2^{2n/p-n}(n!)^{1-1/p}\ . \tag{3.3}$$

Next we turn to m.s.i. with respect to stable noises. The simplest case of symmetric stable, with exponent $\alpha\in(1,2)$ (or shortly, α-stable), noises will be discussed.

For $f:T^n\to\mathbb{C}$, $\alpha>1$, $\varepsilon>0$ such that $\alpha-\varepsilon\ge 1$ set

$$||f||_{\alpha,\varepsilon} = \Big(\sum_D \int_{T^{n-|D|}} \Big(\int_{T^{|D|}} |f(t_1,\dots,t_n)|^{\alpha+\varepsilon} d\mu^{|D|}\Big)^{\frac{\alpha-\varepsilon}{\alpha+\varepsilon}} d\mu^{n-|D|}\Big)^{\frac{1}{\alpha-\varepsilon}}, \tag{3.4}$$

where the sum is taken over all subsets $D\subset\{1,\dots,n\}$, $|D|$ = the number of elements in D, the inner integral is taken with respect to t_j, $j\in D$ and the outer one with respect to the rest of variables. Denote $L^{\alpha,\varepsilon}(T^n)$ the Banach space of functions f with finite norm $||f||_{\alpha,\varepsilon}$. If f is symmetric, $||f||_{\alpha,\varepsilon}$ is equivalent to (0.9). Simple functions $f: T^n\to\mathbb{C}$ are dense in $L^{\alpha,\varepsilon}(T^n)$.

Theorem 3.2. Let Z be stable noise in T with ch.f. (0.8), $\alpha\in(1,2)$ and $\varepsilon>0$ such that $1\le\alpha-\varepsilon<\alpha+\varepsilon\le 2$. For any $f\in L^{\alpha,\varepsilon}(T^n)$ there exists r.v. $I^{(n)}(f) = \int_{T^n} f(t_1,\dots,t_n)Z(dt_1)\dots Z(dt_n)$ defined as the limit in $L^{\alpha-\varepsilon}(\Omega)$ of integral sums $I^{(n)}(f_k) = \sum_{\Delta_1,\dots,\Delta_n} f_k^{\Delta_1,\dots,\Delta_n} Z(\Delta_1)\dots Z(\Delta_n)$, where

$(f_k)_{k\geq 1}$ is a sequence of simple functions convergent to f in $L^{\alpha,\varepsilon}(T^n)$ such that

(j1) $E[|I^{(n)}(f)|^{\alpha-\varepsilon}] \leq C||f||^{\alpha-\varepsilon}_{\alpha,\varepsilon}$, $C=C(n,\alpha,\varepsilon)$,

(j2) $E[I^{(n)}(f)] = 0$,

(j3) $I^{(n)}(f) = I^{(n)}(\text{symf})$.

Proof. There exists Poisson noise $p(dt,du)$ in $T\times\mathbb{R}_0$ ($\mathbb{R}_0 = \mathbb{R}\setminus\{0\}$), defined on the same probability space and measurable with respect to σ-algebra $F = \sigma(Z(A), A\in B_c(T))$ with mean $E[p(dt,du)] = c\mu(dt)du/|u|^{\alpha+1}$ such that

$$Z(A) = \int_{A\times\mathbb{R}_0} uq(dt,du), \quad A\in B_c(T), \tag{3.5}$$

where $q(dt,du)=p(dt,du)-E[p(dt,du)]$, $c=(-2\Gamma(-\alpha)\cos(\pi\alpha/2))^{-1}$. Set $Z_2(A) = \int_{A\times\{|u|>1\}} uq(dt,du)$, $Z_1(A)=Z(A)-Z_2(A)$, hence for simple f,

$$I^{(n)}(f) = \sum_D \sum_{\Delta_j:j\notin D} \Big(\sum_{\Delta_i:i\in D} f^{\Delta_1,\dots,\Delta_n} \prod_{i\in D} Z_1(\Delta_i)\Big) \prod_{j\notin D} Z_2(\Delta_j),$$

where the sum $\sum_D$ is analogical as in (3.4). Z_1 and Z_2 being independent, denote E_j expectation with respect to Z_j, j=1,2. Below C will denote constants depending on n,α,ε but not on f. By Theorem 3.1 and (3.5),

$$E_2[|I^{(n)}(f)|^{\alpha-\varepsilon}]\leq C \sum_D \sum_{\Delta_j:\notin D} \Big|\sum_{\Delta_i:i\in D} f^{\Delta_1,\dots\Delta_n} \prod_{i\in D} Z_1(\Delta_i)\Big|^{\alpha-\varepsilon} \prod_{j\notin D} \mu(\Delta_j),$$

and

$$E|I^{(n)}(f)|^{\alpha-\varepsilon} = E_1E_2|I^{(n)}(f)|^{\alpha-\varepsilon}$$

$$\leq C \sum_D \sum_{\Delta_j:j\notin D} \Big(E_1\Big[\Big|\sum_{\Delta_i:i\in D} f^{\Delta_1,\dots,\Delta_n} \prod_{i\in D} Z_1(\Delta_i)\Big|^{\alpha+\varepsilon}\Big]\Big)^{\frac{\alpha-\varepsilon}{\alpha+\varepsilon}} \prod_{j\notin D} \mu(\Delta_j)\leq C||f||^{\alpha-\varepsilon}_{\alpha,\varepsilon}.$$

The rest of the proof is easy. ∎

As an application of Th.3.2 we shall define a class of self-similar processes with stationary increments. We recall that a (real' -valued) process $X=X(t), t\geq 0$ is said <u>self-similar with index</u> $\chi\in\mathbb{R}$ if for any $\lambda>0$, finite dimensional distributions of X and $X_{\chi,\lambda} = \lambda^{\chi}X(\lambda t)$, $t\geq 0$ coincide [2], [15].

Let $T=\mathbb{R}$, $\mu(dt)=dt$, and let Z be α-stable ($1<\alpha<2$) noise in T with ch.f. (0.8). Introduce stochastic processes

$$(3.6)\quad Z^{(n)}(t) = \int_{\mathbb{R}^n} \left(\int_0^t ds/|x_1-s|^{\beta} \ldots |x_n-s|^{\beta} \right) Z(dx_1)\ldots Z(dx_n),\ t \ge 0.$$

In case of Gaussian noise Z analogical processes called <u>Hermite processes</u> were studied in [15], [16]. Linear processes $Z^{(1)}(t)$ (3.6) were discussed in [1].

<u>Theorem 3.3</u>. Processes $Z^{(n)}$ (3.6) are well-defined for

$$(3.7)\qquad 1/\alpha < \beta < 1/\alpha + (\alpha-1)/\alpha n,\quad n=1,2,\ldots ,$$

they are self-similar with index $\chi=n\beta-1-n\alpha^{-1}$ $(-1<\chi<-1/\alpha)$, have stationary increments and continuous trajectories. Moreover,

$$E[Z^{(n)}(t)]= 0 \quad \text{and} \quad E[|Z^{(n)}(t)|^{\alpha-\varepsilon}] < \infty,\ t\ge 0,\ \ \varepsilon>0.$$

Proof. Denote $f(t,x_1,\ldots,x_n) = \int_0^t ds/|x_1-s|^{\beta}\ldots|x_n-s|^{\beta},\ x_1,\ldots x_n\in\mathbb{R}$. Consider integrals $J_n(t,\alpha,\beta) = ||f(t,\cdot)||^{\alpha}_{L^{\alpha}(\mathbb{R}^n)}$. By generalized Young inequality (e.g. [12], Ch.IX.4), if $\alpha>0$, $0<\beta<1$ and $\alpha\beta<1$, then

$$(3.8)\qquad J_1(t,\alpha,\beta) \le Ct^{\alpha_1}$$

and

$$J_n(t,\alpha,\beta) \le CJ_{n-1}(t,\alpha_1,\beta_1)\ ,\quad n > 1\ ,$$

where $\alpha_1 = 1-\alpha\beta+\alpha$, $\beta_1 = \alpha\beta/\alpha_1$, and C (here and below) is a finite quantity dependent on α, β. Hence if (3.7) holds then

$$(3.9)\qquad J_n(t,\alpha,\beta) \le CJ_1(t,\alpha_{n-1},\beta_{n-1}) \le Ct^{\alpha_n},$$

where $\alpha_{n-1} = (n-1)(1-\alpha\beta) + \alpha$, $\beta_{n-1} = \alpha\beta/\alpha_{n-1}$. Consequently $J_n(t,\alpha,\beta)$ is finite under condition (3.7). Similarly it can be proved that $f(t,.)\in L^{\alpha,\varepsilon}(\mathbb{R}^n)$ for sufficiently small $\varepsilon>0$, i.e. $Z^{(n)}(t)$ is well-defined under (3.7). By (j1) and an estimate analogical to (3.9) it can be shown that $E|Z^{(n)}(t)|^{\alpha-\varepsilon}\le C||f(t,)||^{\alpha-\varepsilon}_{L^{\alpha,\varepsilon}(\mathbb{R})} \le Ct^{1+\delta}$ with some $\delta>0$, hence by Kolmogorov's test processes (3.6) have continuous paths. As noises $Z(dt)$ and $\lambda^{1/\alpha} Z(d\lambda t)$ are identically distributed for any $\lambda > 0$, processes $Z^{(n)}$ are self-similar with χ given in Th. 3.3 (the details of the proof are analogical to the case of Gaussian noise Z [2]). The rest follows from Th.3.2, (j1)-(j2). ∎

Remark 3.4. In the Gaussian case ($\alpha = 2$) and $n \geq 2$, the bounds (3.7) are exact, i.e. the corresponding Hermite processes do not exist if β is equal to extremes sides of (3.7). It seems that the same is true in α-stable case ($1<\alpha<2$) too but rigorous proof would require more deep analysis of m.s.i. with respect to stable noises.

4. Appendix. We present here another proof of inequality (2.11) communicated by A.Grincevičius.

Let $\mathcal{P}_n$ denote the set of discrete probability distributions $P = \sum_{i=1}^{n} p_i \delta_{\{a_i\}}$ on R with zero mean $\int x dP = \sum a_i p_i = 0$. Denote $\mathcal{P}$ the set of all probability distributions on R with mean 0 and finite variance.

Lemma 4.1. It suffices to prove (2.11) for $P \in \mathcal{P}_n$, $n \geq 2$.

Proof: by standard argument.∎

Lemma 4.2. (2.11) is true for $P \in \mathcal{P}_2$.

Proof: by direct calculation. ∎

The crucial induction step is

Lemma 4.3. Let $n \geq 3$. For any $P \in \mathcal{P}_n$ there exist two distributions $P_i \in \mathcal{P}_{n-1}$, $i=1,2$, and $c \in [0,1]$ such that $P = cP_1 + (1-c)P_2$.

Proof. Let $P = \sum p_i \delta_{\{a_i\}} \in \mathcal{P}_n$, where $a_1 < \ldots < a_n$. Set $A_+ \{a_i : a_i > 0\}$, $A_- = \{a_i : a_i \leq 0, i \neq 1\}$. Then $a_1 < 0$ and $A_+ \neq \emptyset$. Without loss of generality we assume that $A_- \neq \emptyset$.

Let us construct the required probability distributions P_1 and P_2 concentrated on $a_1 \cup A_+$ and $A_- \cup A_+$ respectively, as follows:

$$P_1(\{a_i\}) = \begin{cases} p_1/c & , a_i = a_1, \\ (1-(p_1/c))p_i / \sum_{A_+} p_i, & a_i \in A_+ , \end{cases}$$

and

$$P_2(\{a_i\}) = \begin{cases} p_i/(1-c) & , a_i \in A_- , \\ (1- \sum_{A_-} p_i/(1-c)) / \sum_{A_+} p_i, & a_i \in A_+ \end{cases}$$

($\sum_{A_\pm}$ refer to sums over $i=1,\ldots n: a_i \in A_\pm$ correspondingly, where

(4.1) $$p_1 \le c \le 1 - \sum_{A_-} p_i .$$

It is clear that P_1 and P_2 are probability distributions and $P = cP_1 + (1-c)P_2$. It remains to find c satisfying (4.1) such that $\int x dP_i = 0$, $i=1,2$. By definition this is equivalent to

(4.2) $$p_1 a_1/c + (\sum_{A_+} a_i p_i)(1-(p_1/c)) \sum_{A_+} p_i = 0$$

and

(4.3) $$\sum_{A_-} p_i a_i/(1-c) + (\sum_{A_+} a_i p_i)(1- \sum_{A_-} p_i/(1-c))/ \sum_{A_+} p_i = 0.$$

Introduce $\lambda = \sum_{A_+} a_i p_i / \sum_{A_+} p_i$, $\lambda > 0$. By (4.2), $c=c_1=p_1(\lambda-a_1)/\lambda > p_1$ $(-a_1>0)$, and by (4.3), $1-c=1-c_1=-\sum_{A_-} a_i p_i / + \sum_{A_-} p_i \ge \sum_{A_-} p_i$. It remains to verify that $c_1=c_2$, or $c_1-1-c_2 = 1$. ∎

The last step is

<u>Lemma 4.4</u>. Let $P_1, P_2 \in \mathcal{P}$ satisfy (2.11), $c\in[0,1]$. Then $P=cP_1 + (1-c)P_2$ again satisfies (2.11).

Proof. Let φ_i be ch.f. of $P_i \in \mathcal{P}$, $\sigma_i^2=-\varphi_i''(0)$, $i=1,2$. Then $\varphi = \int e^{iax} dP$ $= c\varphi_1 + (1-c)\varphi_2$, $\sigma^2 = \int x^2 dP = c\sigma_1^2 + (1-c)\sigma_2^2$. By the Cauchy inequality,

(4.4) $$|\varphi|^2 = |c\varphi_1 + (1-c)\varphi_2|^2 \le c|\varphi_1|^2 + (1-c)|\varphi_2|^2$$

and

$$|\varphi'|^2/\sigma^2 = |c \frac{\varphi_1'}{\sigma_1} \sigma_1 + (1-c) \frac{\varphi_2'}{\sigma_2} \sigma_2|^2/\sigma^2 \le c|\varphi_1'|^2/\sigma_1^2 + (1-c)|\varphi_2'|^2/\sigma_2^2 .$$

(4.5)

Summing up (4.4)-(4.5) one has (2.11). ∎

References

[1] Astrauskas, A.: On stable self-similar fields (Russian). Liet. matem. rink. (to appear).

[2] Dobrushin, R.L.: Gaussian and their subordinated self-similar random generalized fields. Ann. Probab. 7, (1979) 1-28.

[3] Hida, T.: Stationary stochastic processes. Princeton, N.J.: Princeton University Press 1970.

[4] Ito, K.: Multiple Wiener integral. J. Math. Soc. Japan 3, (1951) 157-164.

[5] Ito, K.: Spectral type of shift transformations of differential process with stationary increments. Trans. Amer. Math. Soc. 81, (1956) 253-263.

[6] Kabanov, Ju.M.: On extended stochastic integrals (Russian). Teor. Verojatn. i Primen. 20, (1975) 725-737

[7] Kallianpur, G.: The role of reproducing kernel Hilbert spaces in the study of Gaussian processes. In: Hida T. (ed.). Advances in Probability and Related Topics, Vol. 2, p. 49-83. N.Y.: Dekker 1970.

[8] Lass Fernandez, D.: Lorentz spaces, with mixed norms. J. Funct. Anal. 25, (1977) 125-146.

[9] Meyer, P.-A.: Un cours sur les integrales stochastiques. Seminaire de Probabilites X. Lect. Notes Math. Vol. 511. Springer 1976.

[10] Ogura, H.: Orthogonal functionals of the Poisson process. IEEE Trans. Inform. Theory IT-18, (1972) 473-480.

[11] Okazaki, Y.: Wiener integral by stable random measures. Memoir Fac. Sci. Kyushu Univ. Ser. A, 33, (1979) 1-70.

[12] Reed, M., Simon, B.: Methods of modern mathematical physics: Fourier analysis, self-adjointness. N.Y.: Academic Press 1975.

[13] Surgailis, D.: On multiple Poisson stochastic integrals and associated Markov semigroups (to appear).

[14] Surgailis, D.: On infinitely divisible self-similar random fields (to appear).

[15] Taqqu, M.S.: Self-similar processes and related ultraviolet and infrared catastrophes. Techn. report No. 423 Cornell Univ. 1979.

[16] Taqqu, M.S.: A representation for self-similar processes. Stoch. Processes Appl. 7, (1978) 55-64.

[17] Wiener, N.: The homogeneous chaos. Amer. J. Math. 60, (1938) 897-936.

Inst. of Mathematics and Cybernetics
232600, Vilnius, 54.
K.Pozelos st.

OPTIMAL STOCHASTIC CONTROL UNDER RELIABILITY CONSTRAINTS

D. Vermes

Bolyai Institute
University of Szeged
Hungary

Abstract

The expected discounted reward connected with the controlled evolution of a uniformly nondegenerate n-dimensional diffusion process is to be maximized under the additional constraint that the process may leave a fixed set of its state space only with probability not greater than ε. The paper contains an appropriate definition of constrained optimality and a constructive proof of the existence of an optimal control strategy.

Introduction

In many real-life control problems the objective of the controller cannot be reduced to the maximization of a single performance functional. Dealing with a chemical reactor one generally aims an optimal performance under usual working conditions. But controlling the plant one may not forget that under unfortunate coincidence of disturbing factors the whole reactor can explode. The situation is similar at the control of a dam used to provide water for a power station or for watering the fields, when the danger of a flood-disaster arises. The costs of the consequences of a catastrophe are not comparable with the reward of the usual operation of the plant, hence the both cannot be united to a single performance functional.

The problem can be formulated as follows. Maximize the reward resulting from the ordinary operation of the system, but use only such control strategies for which the probability of the occurance of the catastrophe is less than a given ε.

This problem can be regarded as the control theoretic (dynamic) counterpart of probability constrained stochastic programming (c.f. [3]).

The constrained control problem

To focus attention to the most essential new features of constrained problems we consider the optimal control of diffusion processes which is the most extensively studied example in the non-constrained case (c.f. [1], [2]). Let the underlying processes x_t be R^n valued diffusions with uniformly elliptic diffusion matrix $a(x)$ and controllable drift vector $b(x,d)$. The decisions d can be choosen from a compact set $D \subset R^m$ according to a measurable function (a Markov strategy) $u(x)$ depending on the state x only. We denote by E_x^u, P_x^u the conditional expectation and probability corresponding to the process controlled by strategy u and starting from state x. (We drop upper indicies u of x and σ.) A catastrophe occures if the state of the process leaves a fixed set $\Gamma \subset R^n$. Hence the random time σ when the disaster happens is the first exit time of the process from Γ. Our objective is to choose a strategy u which maximizes the discounted reward

$$(1) \qquad J_x(u) = E_x^u \int_0^\sigma e^{-\alpha t} q(x_t,u(x_t)) \, dt$$

and for which the reliability constraint

$$(2) \qquad R_x(u) = P_x^u (\{x_t \in \Gamma, \quad \forall t\}) \geq \rho = 1-\varepsilon$$

is satisfied. Here the discount rate $\alpha > 0$ and the reward rate function $q(x,d)$ are given. For technical reasons we assume for the whole paper that functions a,b,q and open set Γ are bounded and lipschitzian.

It is well known that discounting the reward is equivalent with killing the process at an independent, α-exponentially distributed termination time ζ. We shall make use of this possibility. Henceforth E_x^u and P_x^u will always refer to the probabilities connected with the terminating processes, while we use $\tilde{P}$, $\tilde{E}$ for the non-terminating processes. This enables us to drop factor $e^{-\alpha t}$ from (1) and to rewrite (2) in the form

$$(2') \qquad R_x(u) = \int_0^\infty \tilde{P}_x^u (\sigma>t)\alpha e^{-\alpha t} \, dt = \alpha \int_0^\infty \tilde{P}_x^u (\sigma>t,\zeta>t) \, dt = \alpha E_x^u \sigma \geq \rho$$

The constrained optimal control problem is not yet well-posed until we have not specified the initial state. The condition that the optimal strategy should maximize the reward for one fixed initial point x_o only, i.e. $J_{x_o}(u^*) = \min\{J_{x_o}(u): R_{x_o}(u) \geq \rho\}$ is too mild, it would result too many optimal strategies. The condition used in the classical unconstrained case, that u should be optimal for any initial state (i.e. $J_x(u) = \min\{J_x(u): R_x(u) \geq \rho\}$ $\forall x$) is too strong as there is no strategy satisfying $R_x(u) \geq \rho$ for all x unless $\rho = 0$.

(There does not even exists a strategy maximizing the reward for all x and satisfying reliability constraint for one fixed x_o. That is the reason why we have changed our approach since the preliminary abstract.)

In order to give sense the word "optimal strategy" for constrained control problems let us define for each reliability level ρ and for each strategy u the <u>optimality set</u>

$$(3) \qquad \Omega_\rho(u) = \{x : J_x(u) = \min\{J_x(v): R_x(v) \geq \rho\}\}.$$

<u>Definition 1</u>. At reliability level ρ we call a strategy u_1 better than u_2 if

$$\Omega_\rho(u_1) \supset \Omega_\rho(u_2)$$

If there is a maximal element u* with respect to this ordering, then we call it the <u>optimal strategy at reliability level</u> ρ. Notice that for $\rho = 0$ our definition reduces to the classical unconstrained one.

<u>Related unconstrained problems</u>

The theory of non-constrained optimal control of diffusion processes is well developed (c.f. [1], [2]). Here we list the most important results of this theory which will serve as basis for further reference and comparision. Under the above conditions on a, b, q, and Γ the following statements are valid.

(a). There exists an optimal Markovian strategy

(b). None of the strategies using information about the past history of the processes is better than the optimal Markov strategy.

(c). The value function $\psi(x) = \inf_u J_x(u)$ is twice continuously differentiable, and it is the unique C_2-solution with boundary values $\psi(x) = 0$ on $\partial\Gamma$ satisfying the Hamilton-Jacobi-Bellman equation

$$(4) \qquad \sum_{i,j} a_{ij}(x) \frac{\partial^2\psi(x)}{\partial x_i \partial x_j} - \rho\psi(x) + \inf_{d\in D} \{\sum_i b_i(x,d) \frac{\partial\phi(x)}{\partial x} + q(x,d)\} = 0$$

in Γ.

There are two special cases of our constrained control problem for which these classical results are valid.

(A). If we set $\rho = 0$ we only have to maximize (1), which is a classical non-constrained <u>maximal reward problem</u>. Let u_A denote its optimal strategy. It follows from Definition 1 that on the set $\{x: \rho \le R_x(u_A)\}$ the optimal strategy u^* should not be worse than u_A. (It is possible that the optimal u_A is not unique, and to different optimal strategies there belong different reliability functions R_x. In this case we denote by u_A that particular optimal policy, for which the reliability is maximal among all optimal strategies. The existence of such an u_A follows from (4).)

(B). If we set $q(x,d) = 1/\alpha$ and $\rho = 0$, then we get again a classical non-constrained control problem, which we call the <u>maximal reliability problem</u>. This name is justified by the fact that by the special choice of q for any strategy u we have $J_x(u) = R_x(u)$. Hence to the optimal strategy u_B of this problem there belongs the value function $\sup_u R_x(u)$. In the case of non-uniqueness we denote by u_B a strategy resulting the largest reward $J_x(u_B) = E_x^{u_B} \int_\rho^\sigma q(x_t, u(x_t))dt$ among all policies optimal from point of view of maximal reliability. One can see immediately that for those states x for which $\rho > \sup_u R_x(u)$ there exists no strategy satisfying the reliability constraint. Hence at reliability level ρ the maximal possible optimality set is

$$\Omega_\rho = \bigcup_u \Omega_\rho(u) = \{x: \sup_u R_x(u) \ge \rho\}.$$

Piecewise Markovian strategies

As it can be seen from statements (a) and (b) of the previous paragraph the class U_M of Markovian strategies is sufficiently broad for the purposes of unconstrained control problems. But the situation is different for constrained problems. Form (2') of the reliability constraint shows clearly that the time t spent since the beginning of the process contains essential information for the controller. E.g. if in time interval $[0,\rho/\alpha)$ the catastrophe did not happen then after $t = \rho/\alpha$ any control strategy can be applied, the reliability constraint is satisfied with absolute certainty. This fact forces us to introduce a broader class of state and time dependent strategies.

Definition 2. The class U of functions u is called the class of piecewise Markovian strategies if $U_M \subset U$ and if for any stopping time τ with u_1 and u_2 also

$$u(t,x)=\begin{cases} u_1(t,x) & \text{if} \quad t \le \tau \\ u_2(t,x) & \text{if} \quad t > \tau \end{cases}$$

belongs to U.

As stopping times are measurable with respect to the past of the processes, piecewise Markovian strategies are non-anticipating. Notice that the controlled process corresponding to the strategy $u \in U \backslash U_M$ is not necesserily Markovian. But if $u_1, u_2 \in U_M$ and u is defined as in Definition 2, then the conditioned measures $P_x^u\ (\cdot\ |t \le \tau)$ and $P_x^u\ (\cdot\ |t > \tau)$ describe Markov processes. Esspecially

$$E_x^u f(x_t) = E_x^{u_1} [\chi_{t \le \tau}\ f(x_t) + \chi_{t > \tau}\ E_{x(\tau)}^{u_2}\ f(x_{t-\tau})]$$

holds true for any bounded function f. This property, which we call piecewise Markov property remains valid also for the case of several switching times $\tau_1, \tau_2, \ldots \tau_N$.

Structure of the optimal strategy

Theorem. If $\rho \le \sup_x \sup_u R_x(u)$ then there exists an optimal piecewise Markovian strategy u* with the maximal possible optimality set

$$\Omega_\rho(u^*) = \Omega_\rho = \{x : \rho \le \sup_u R_x(u)\}.$$

Recall that u_A and u_B denote the optimal strategies for the maximal reward problem (A) and the maximal reliability problem (B) respectively. Let us introduce the notation

$$A = \{(t,x) : \rho - \alpha t \le R_x(u_A)\} \quad \text{and} \quad B = \{(t,x) : \rho - \alpha t > R_x(u_B)\}$$

and denote τ_A and τ_B the first entrance times into the sets A and B respectively. If we define the function

$$p(t,x) = \begin{cases} J_x(u_A) & \text{if} \quad (t,x) \in A \\ J_x(u_B) & \text{if} \quad (t,x) \in B \end{cases}$$

on $A \cup B$ and denote $\tau = \min(\tau_A, \tau_B)$ then we can formulate a new unconstrained optimal control problem (C) on the set $C = [0,\infty)\times\Gamma\setminus(A\cup B)$:

$$\text{maximize } E^u_{t,x} \{ \int_0^\tau q(x_s, u(s,x_s))\, ds + p(\tau, x_\tau)\} \quad \text{for all} \quad (t,x)\in C$$

By statement (a) there exists an optimal Markovian strategy u_C solving this problem.

<u>At reliability level ρ the optimal piecewise Markovian strategy u^* of the constrained control problem has the following form</u>

$$u^*(t,x) = \begin{cases} u_A(x) & \text{if} \quad t \geq \tau = \tau_A, \\ u_B(x) & \text{if} \quad t \geq \tau = \tau_B, \\ u_C(t,x) & \text{if} \quad t < \tau. \end{cases}$$

<u>Justification of the optimality</u>

The points $(t,x)\in A$ are characterized by the property that the time spent since the beginning is larger than $(\rho - R_x(u_A))/\alpha$. Observe that

$$\alpha \cdot E^u_{0,x_0} (\sigma | x_t = x \in A,\ \sigma > t) = \alpha \cdot t + R_x(u_A) \geq \rho$$

for any initial point x_0 and for any $u\in U$ which coincides with u_A after t. Consequently if for $(t,x)\in A$ one applies u_A from time t on then the reliability constraint need not be taken into consideration, it is automatically satisfied. But as u_A is optimal not only in U_M but also in U for the non-constrained maximum reward problem, and it does not hurt the reliability constraint, it is also optimal for the constrained problem from any $(t,x)\in A$ on.

Observe that using u_A and u_B after τ, we get the reliability values $R_{t,x}(u^*) = \rho - \alpha t$ at any boundary point (t,x) of C. Hence by the piecewise Markov property

$$R_{0,x}(u) = E^u_{0,x}(\alpha \cdot \tau + R_{\tau, x(\tau)}(u^*)) = \rho$$

for any initial point $(0,x)\in C$ and for any strategy u coinciding with u^* after τ. We have already seen that after τ_A one has necessarily to apply u_A. Let u be as just specified and v an arbitrary strategy coinciding with u everywhere but on $\{t > \tau = \tau_B\}$. Then from the continuity of $R_{t,x}(v)$ and from the maximality of $R_{t,x}(u_B)$ on B it follows that there exists an open set G on the boundary of C such that $R_{t,x}(v) < R_{t,x}(u) = R_{t,x}(u^*)$ on $(t,x)\in G$. From the non-degenerateness of the processes follows that the open set G can be reached with positive $P^u_{0,x}$-probability from every x, and we obtain

$$R_{0,x}(v) = E^v_{0,x}\ (\alpha\cdot\tau + R_{\tau,x(\tau)}\ (v))=$$

$$= E^u_{0,x}\ (\alpha\cdot\tau + R_{\tau,x(\tau)}\ (v)) < R_{0,x}\ (u)$$

for every initial point $(0,x)\in C$. From the maximality of $R_{t,x}(u^*)$ on B follows that after τ it is impossible to compensate the reliability deficit caused by v on G. As v was arbitrary before τ we see that any strategy coinciding with u_A on $\{t > \tau = \tau_A\}$ and deviating from u_B on $\{t > \tau = \tau_A\}$ results a reliability $R_{0,x}(v) < \rho$ for every starting point $(0,x)\in C$.

We have still to prove that the set C is choosen optimally and that u* is optimal on C. As we have just seen after the first exit τ from C the optimal strategy is necessarily u*. Hence the optimal C* can be only smaller than the one defined in the previous paragraph. As from the construction of C follows that the reliability constraint is automatically satisfied for every initial point $(t,x)\in C$, the control problem (C) can be regarded as non-constrained. Suppose C* is smaller than C. This would imply that optimally one should switch already at the first exit τ^* from C , i.e. before τ. This would mean that the optimal strategy for (C) is non-Markovian. But as it cannot be better than the optimal Markovian strategy u_C we can disregard the effect of an eventually smaller C*.

By the piecewise Markov property we have

$$J_{0,x}(u^*) = \max_{v\in U}\ E^v_{0,x}\ \{\ \int_0^\tau q(x_t,\ v(t,x_t))\ dt\ +$$

$$+ \chi_{\tau=\tau_A}\ \max_{u\in U} E^u_{\tau,x(\tau)} \int_0^\sigma q(x_t,\ u(x_t))\ dt\ + \tag{5}$$

$$+ \chi_{\tau=\tau_B}\ E^{u_B}_{\tau,x(\tau)} \int_0^\sigma q(x_t,\ u_B(x_t))\ dt\}$$

In this formula a change in u_B and τ would either brack the reliability constraint or give no gain in J. As otherwise both maximizations are free and the constraint is satisfied (5) shows the optimality of u^*.

Acknowledgement

This work was done while the author was with the Operations Research Department of the Computer and Automation Institute of the Hungarian Academy of Sciences. The author is indebted to the head of the Department Professor A.Prékopa for encouragement and motivating discussions.

REFERENCES

[1] Fleming, W.H. and Richel, R.: Deterministic and Stochastic Optimal Control. Springer-Verlag, 1975.

[2] Krylov, N.V.: Control of Diffusion Processes /in Russian/. Moscow, 1977.

[3] Prékopa, A.: On probabilistic constrained programming, in Proceedings of the Princeton Symposium on Mathematical Programming, pp. 113-138. Princeton, 1970.

Bolyai Institute, University of Szeged
6720 Szeged, Aradi vértanuk tere 1.
Hungary

ON CONTROLLED SEMI-MARKOV PROCESSES WITH AVERAGE REWARD CRITERION

A.A. Yuškevič
Moscow

Our aim is to show that if a stationary policy φ satisfies average optimality equations in a controlled semi-Markov model with Borel state and action spaces, then φ is optimal in the sense that

(1) $$\overline{\lim}\ t^{-1}w(x,\pi,t) \le \lim t^{-1}w(x,\varphi,t) \quad \text{as } t \to \infty,$$

where π is an arbitrary policy and $w(x,\pi,t)$ denotes the expected reward up to time t under the initial state x and policy π.

In controlled discrete-time Markov processes this is a well known result. For semi-Markov models with finite state and action spaces it is known that average optimality equations imply optimality of φ in the sense that

(2) $$\lim t^{-1}w(x,\psi,t) \le \lim t^{-1}w(x,\varphi,t) \quad \text{as } t \to \infty,$$

(3) $$\underline{\lim}\ t^{-1}w(x,\pi,t) \le \lim t^{-1}w(x,\varphi,t) \quad \text{as } t \to \infty,$$

π being arbitrary and ψ arbitrary stationary policies ([1] - [4]). For infinite Borel controlled semi-Markov processes the result was obtained only in the case when $\lim t^{-1}\ w(x,\varphi,t)$ does not depend on x in form

(4) $$\overline{\lim}\ T_n^{-1}(x,\pi)w_n(x,\pi) \le \lim T_n^{-1}(x,\varphi)w_n(x,\varphi), \text{ as } n \to \infty,$$

where $T_n(x,\pi)$ is the expected time of n-th renewal and $w_n(x,\pi)$ is the expected reward up to the n-th renewal ([5], [6], [7]). Under some additional recurrency assumptions,(4) implies (2), but whether (4) implies more natural criteria (1) or (3) is not clear([5], [6]). We consider directly criterion (1).

The model in question is defined by following elements: 1) A state space X, 2) an action space A, 3) a projection $j:A \to X$, where $A(x) = j^{-1}(x)$ is the set of actions admissible at state x, 4) a transition function $P_a(d\tau d\xi)$ from A to $(0,\infty) \times X$, 5) a reward function $r(a,t)$ on $A \times [0,\infty)$. All the elements are Borel measurable, the correspondence $x \to A(x)$, $x \in X$ admits a measurable selector.

Let $x_o a_1 t_1 x_1 a_2 t_2 \ldots$ be successively attended states, chosen actions and renewal moments, with $t_o = 0$. A policy π is a collection of conditional distributions $\pi_n(d\, a_{n+1} | x_o a_1 \ldots t_n x_n)$ concentrated on $A(x_n)$, $n = 0,1,2,\ldots$. Conditional distribution of $(t_{n+1} - t_n, x_{n+1})$ given $x_o a_1 \ldots t_n x_n a_{n+1}$ is $P_{a_{n+1}}$. So the choice of an initial state $x_o = x$ and a policy π defines the probability measure of the process, and by E_x^π we designate the corresponding expectation. We let

$$(5) \qquad w(x,\pi,t) = E_x^\pi \left[\sum_{n=1}^{N} r(a_n, t_n - t_{n-1}) + r(a_{N+1}, t - t_N) \right]$$

with $N = \max(n : t_n \le t)$ (we don't write argument x in r because as in [8] our element a_{n+1} is equivalent to usual pair $x_n a_{n+1}$). A policy is called stationary and denoted φ instead of π, if $\pi_n(\varphi(x_n) | x_o \ldots a_n t_n x_n) = 1$, $n = 0,1,2,\ldots$ for some measurable selector $a = \varphi(x)$ of the correspondence $x \to A(x)$. For brevity by τ and ξ we denote a pair of random variables with joint distribution $P_a(d\tau d\xi)$ depending on a.

<u>Theorem</u>. Suppose there exist a number $\varepsilon > 0$ and a decreasing finite function $f(t)$ with $f(+\infty) = 0$ such that uniformly in a for all $t \ge 0$

(i) $P_a\{\tau > \varepsilon\} > \varepsilon$;

(ii) $E_a I_{\{\tau \ge t\}} \tau \le f(t)$;

(iii) $E_a I_{\{\tau \ge t\}} [|r(a,\tau)| + |r(a,t)|] \le f(t)$.

Define $r(a) = E_a r(a,\tau)$. If a stationary policy φ and bounded measurable functions g and h on X satisfy the average optimality equations

$$(6) \qquad \begin{aligned} g(x) &= E_{\varphi(x)} g(\xi) = \max_{a \in A(x)} E_a\, g(\xi), \\ h(x) &= r(\varphi(x)) + E_{\varphi(x)}[h(\xi) - g(\xi)\tau] = \max_{a \in A(x)} E_a[r(a) + h(\xi) - g(\xi)\tau], \end{aligned}$$

then (1) holds for all x and π.

From previously known results it follows that if X and A are finite than the system (6) has a solution. Conditions (i) - (iii) in this case reduce to $0 < E_a \tau < \infty$, $E_a |r(a,\tau)| < \infty$, $\lim E_a I_{\{\tau \ge t\}}\, r(a,t) = 0$.

The proof of the theorem is based on the semi-Markovian analogue of the concept of "canonical policies" (cf. [8]). Instead of (5) define

$$(7) \qquad W(x,\pi,t) = E_x^\pi \sum_{n=1}^{N} r(a_n) + R(a_{N+1}, t-t_N)$$

where $r(a) = E_a r(a,\tau)$ and

$$R(a,t) = \begin{cases} P_a\{\tau > t\}^{-1} E_a 1_{\{\tau > t\}}[h(\xi)-g(\xi)(\tau-t)] & \text{if } P_a\{\tau > t\} > 0, \\ 0 & \text{if } P_a\{\tau > t\} = 0. \end{cases}$$

<u>Lemma</u>. If the conditions (i) and (ii) and the equations (6) are fulfilled, then for all x,π and $t \geq 0$

$$(8) \qquad W(x,\pi,t) \leq tg(x) + h(x) = W(x,\varphi,t).$$

This can be verified by direct substitution of $tg(x) + h(x)$ into the Bellman equation for the problem with horizon t and final reward R, using the uniqueness property of a bounded solution. From (8) it follows that W satisfies (1). Condition (iii) permits to substitute in (1) W by w.

References

[1] Jewell W.S.Markov renewal programming I and II, Oper. Research, 1963, II, 938-971.

[2] Fox B. Markov renewal programming by linear fractional programming, SIAM J.Appl. Math., 1966, 14, 1418-1432.

[3] Denardo E.V., Fox B.L. Multichain Markov renewal programs, SIAM J. APPL. Math., 1968, 16, 468-487.

[4] Romanovskiĭ I.V. The turnpike theorem for semi-Markov decision processes, Proc. Steklov Inst. Math., 1970, III, 249-267.

[5] Ross S.M. Average cost semi-Markov decision processes, J. Appl. Prob., 1970, 7, 649-656.

[6] Hausmann U.G. On the optimal long-run control of Markov renewal processes, J.Math.Anal.Appl., 1971, 36, 123-140.

[7] Federgruen A.,Tijms H.C. The optimality equation in average cost denumerable state semi-Markov decision problems, recurrency conditions and algorithms, J.Appl. Prob., 1978, 15, 356-373.

[8] Dynkin E.B., Yushkevich A.A. Controlled Markov processes Springer-Verlag, New-York, 1979.

A.A. Yuskevic

Moscow Institute of Railway Transport
117526. Moscow
ul.26.Bakinskih kommisarov 10. K2 kv. 104.
U.S.S.R.

LIKELIHOOD RATIOS AND KALMAN FILTERING FOR RANDOM FIELDS

A. V. Balakrishnan
Department of System Science
School of Engineering
University of California, Los Angeles
Los Angeles, CA 90024

Abstract

An exact formula for likelihood-ratios for random fields is developed as well as a Kalman-filter for stationary fields with the correct state space (infinite dimensional).

Introduction

There are many inference problems of recent interest where the observed data is a function of one or more "spatial" parameters, and time is not necessarily a parameter: for instance geophysical data, such as gravity-anomaly or bathymetry. It would appear that the theory of "random fields," (stochastic processes with more than one independent parameter) could contribute much in these problem areas. An informative review paper on statistical geodesy [1] provides a good introduction and an extensive bibliography of the more "applied" literature. The mathematical literature on random fields (see [2,3,4,]) would appear to be polarized on the rigorous mathematical side with concerns far removed from the practical (such as the preoccupation with the concept of Markovianess and Martingales). On the other hand are "engineering" approximations [5,6] which have little mathematical basis.

Of prime importance, as in the case of one-parameter, in inference problems is the likelihood-ratio (of signal-plus-noise to noise-alone) and in this paper we develop an exact formula for the likelihood-ratio in contrast to the heuristic approximate version in [5]. This formula is based on the non-linear white-noise theory [7] of the author and exploits the Krein-Gohberg factorization theory [8].

As again in the one-parameter case, one way to instrument the likelihood-ratio formula is by the Kalman filter. Attempts at deriving Kalman filters for random fields (in the image-processing area) have been made before: see [6] and the references therein. These can at best be described as approximations. The important point is that the state-space for any exact formulation has to be infinite-dimensional as we indicate here.

2. Data Model

Our basic model for the observed data is:

$$v(t_1,t_2) = s(t_1,t_2) + N(t_1,t_2) , \qquad (t_1,t_2) \in \mathcal{D} \subset R^2$$

where $\mathcal{D}$ is a rectangle $0 \le t_1 \le T_1$, $0 \le t_2 \le T_2$. $s(t_1,t_2)$ is the "signal" process; it is assumed to be a Gaussian stochastic process, with zero mean, such that

$$\int_0^{T_2} \int_0^{T_1} E(s(t_1,t_2)^2)\, dt_1\, dt_2 < \infty \tag{2.1}$$

E denoting expectation. We may take the sample space therefore to be $L_2(\mathcal{D})$. Let μ_s denote the countably additive measure induced on $\mathcal{B}$ the Borel sigma-algebra of $L_2(\mathcal{D})$. Then the corresponding characteristic function is

$$\exp - \tfrac{1}{2}[R_s h,h] , \qquad h \in \mathcal{L}_2(\mathcal{D})$$

where [,] denotes inner product in $\mathcal{L}_2(\mathcal{D})$, and R_s is the covariance operator on $\mathcal{L}_2(\mathcal{D})$ into itself, defined by

$$R_s f = g \; ; \quad g(t_1,t_2) = \int_0^{T_1} \int_0^{T_1} R_s(t_1,t_2;\sigma_1,\sigma_2)\, f(\sigma_1,\sigma_2)\, d\sigma_1\, d\sigma_2$$

where

$$R_s(t_1,t_2;\sigma_1,\sigma_2) = E(s(t_1,t_2)\, s(\sigma_1,\sigma_2))$$

and by virtue of (2.1), R_s is nuclear.

The process $N(t_1,t_2)$ is the "noise," idealizing measurement errors. We assume it is white Gaussian. Rather than the usual definition in the "generalized" sense [cf. Gelfand [9]], we employ the "weak-distribution" theory, since we need to go beyond the usual linear filtering problem and have to define non-linear operations. Thus in our "non-linear white noise" theory [10], we take the sample paths to be in $\mathcal{L}_2(\mathcal{D})$ and define the characteristic function to be

$$\exp - \tfrac{1}{2}[h,h] , \qquad h \in L_2(\mathcal{D}) .$$

This characteristic function, as is well-known, defines a weak distribution, or a finitely additive (Gauss) measure μ_G on the algebra of cylinder sets with Borel bases [see [11] for more on this and related notions]. Note that if h_1,h_2 are any two elements in $L_2(\mathcal{D})$, then

$$[N,h_1] \quad \text{and} \quad [N,h_2]$$

are Gaussian distributed, with covariance

$$[h_1,h_2] ,$$

where N denotes the noise sample-function $N(t_1,t_2)$. The main difference brought in by the absence of countable additivity of the measure μ_G is that not every Borel measurable functional is necessarily a random variable [see [11] for more on this]. A measurable function $g(\cdot)$ is called a "physical random variable" if and only if, for any sequence $\{P_n\}$ of finite-dimensional projections converging strongly to the identity, the sequence $\{\phi(P_n,\cdot)\}$ is Cauchy in probability, and all such sequences are equivalent.

We assume further that signal and noise are independent. Hence the characteristic function of the observed process $v(\cdot,\cdot)$ is given by

$$\exp - \tfrac{1}{2}[(I + R_s)h, h] , \qquad h \in L_2(\mathcal{D}) .$$

Let μ_v denote the corresponding weak distribution (finitely additive measure). Then we can prove that μ_v is continuous with respect to μ_G; and further the Radon-Nikodym derivative is a physical random variable and is defined by

$$\phi(v) = \int_{L_2(\mathcal{D})} \exp - \tfrac{1}{2}([s,s] - 2[v,s]) \, d\mu_s \tag{2.2}$$

3. Likelihood-Ratio

We can show that (2.2) defines a physical random variable [see [7]], and is then defined to be the likelihood-ratio of "signal-plus-noise" to "noise" alone. Moreover we can calculate the integral in (2.2) in one of two ways. The first way is to invoke the Krein-Gohberg factorization theory. For this purpose we exploit now the fact that $\mathcal{D}$ is a rectangle. Hence

$$L_2(\mathcal{D}) = L_2[(0,T_1) \; ; \; L_2(0,T_2)] ,$$

in other words we may consider $L_2(D)$ as an L_2-space over the Hilbert Space $L_2(0,T_2)$. Hence we can rewrite our data model

$$v(t,\cdot) = s(t,\cdot) + N(t,\cdot) \qquad 0 < t < T_1$$

where for each t, $s(t,\cdot)$ denotes $s(t,t_2)$ as a function of t_2, $0 < t_2 < T_2$, as an element of $L_2[0,T_2]$ and $s(t,\cdot)$ is Bochner-measurable in t, $0 < t < T_1$. It is convenient to refer to $L_2[(0,T_1) ; L_2(0,T_2)]$ as $W(T_1)$. We note that $N(t,\cdot)$ defines white-noise in $W(T_2)$ and $s(t,\cdot)$ defines a Gaussian process in $W(T_1)$ with covariance operator:

$$[R_s f,g] = E\left\{\int_0^{T_1} [f(t),s(t,\cdot)]dt \cdot \int_0^{T_1} [g(t),s(t,\cdot)]\, dt\right\}$$

where $f(\cdot)$, $g(\cdot) \in W(T_1)$. R_s is a nuclear "covariance" operator and hence we can apply the results in [7] for the more general case where $L_2(T_2)$ is replaced by any separable Hilbert Space $\mathscr{H}$. First, by the Krein-Gohberg theorem:

$$(I + R_s)^{-1} = (I - \mathscr{L}^*)(I - \mathscr{L})$$

where $\mathscr{L}$ is a Hilbert-Schmidt Volterra operator on $W(T_1)$ into $W(T_1)$ defined by:

$$\mathscr{L}f = g \; ; \quad g(t) = \int_0^t L(t,s)\, f(s)\, ds\, , \quad 0 < t < T_1$$
$$f \in W(T_1)\, ,$$

and $(\mathscr{L} + \mathscr{L}^*)$ is nuclear. Moreover, we can as a result, express the likelihood-ratio (2.2) as:

$$\phi(v) = \exp - \frac{1}{2}\left\{[\mathscr{L}v,\mathscr{L}v] - 2[\mathscr{L}v,v] + \int_0^{T_1} \mathrm{Tr}\, L(t,t)\, dt\right\} \qquad (3.1)$$

where the inner-products are in $W(T_1)$, and we have exploited the fact that

$$\int_0^{T_1} \operatorname{Tr} L(t,t)\, dt = \operatorname{Tr} (\mathscr{L} + \mathscr{L}^*) .$$

This then is our likelihood-ratio formula, expressed explicitly as a functional on $W(T_1)$. Note that $\nu = v - \mathscr{L}v$ defines white noise in $W(T_1)$.

To find $\mathscr{L}$, we can follow the explicit construction used by Krein-Gohberg [8]. An alternate technique (as in the one-parameter case) is to invoke the Kalman filtering theory.

4. Kalman Filtering Theory

To apply the Kalman filtering theory we shall consider the "sampled" or "discretized" version as we would need to in any digital computer processing. However we shall discretize only along one direction. Thus let

$$s(t,\cdot) = \text{col } s(t,k\Delta), \qquad k = 0, 1, 2, \ldots, (m-1), \tag{4.1}$$

and

$$m\Delta = T_2 ; \qquad 0 < t < T_1 .$$

Then (4.1) defines an $m \times 1$ multidimensional stochastic process. The Hilbert Space $L_2(T_2)$ is now replaced by R^m.

Second, we assume that the process $s(t,\cdot)$ is stationary in t:

$$E[s(t,\cdot)\, s(\sigma,\cdot)^*] = R(t - \sigma)$$

where $R(\tau)$ is of course $m \times m$ matrix function, and

$$R(-\tau) = R(\tau)^* .$$

Further, we assume that the process has a spectral density: that

$$R(\tau) = \int_{-\infty}^{\infty} e^{i\lambda\tau}\, \phi(\lambda)\, d\lambda$$

where, moreover, the spectral density matrix is such that it is non-singular for every λ, and

$$\int_{-\infty}^{\infty} \frac{\log |\Phi(\lambda)|}{1+\lambda^2}\, d\lambda \quad > \quad -\infty \tag{4.2}$$

where $|\cdot|$ denotes determinant.

The question that arises immediately is: when does a stationary random field $s(t_1,t_2)$ satisfy the condition (4.2)?

<u>Theorem</u>. Suppose that the random field $s(t_1,t_2)$, $(t_1,t_2) \in R^2$, is stationary with spectral density $p(\lambda_1,\lambda_2)$. That is:

$$R_2(\tau_1,\tau_2) \quad = \quad \int_{-\infty}^{\infty}\int_{-\infty}^{\infty} e^{2\pi i(\lambda_1\tau_1+\lambda_2\tau_2)}\, p(\lambda_1,\lambda_2)\, d\lambda_1\, d\lambda_2$$

where

$$R_2(\tau_1,\tau_2) \quad = \quad E[s(t_1+\tau_1,\, t_2+\tau_2)\, s(t_1,t_2)].$$

Suppose that

$$\int_{-\infty}^{\infty} \frac{\log \mu(\lambda)}{1+\lambda^2}\, d\lambda \quad > \quad -\infty \tag{4.3}$$

where

$$\mu(\lambda) \quad = \quad \inf_{\frac{-1}{2\Delta}\le\gamma\le\frac{1}{2\Delta}} \quad \sum_{-\infty}^{\infty} p(\lambda,\gamma+\tfrac{k}{\Delta}) \; . \tag{4.4}$$

Then the condition (4.2) is satisfied by the (one-parameter multi-dimensional) process col. $s(t,k\Delta)$, $k = 0, 1, \ldots, m-1$.

<u>Proof</u>. We shall show first that

$$[\Phi(\lambda)a,a] \;\ge\; \mu(\lambda)[a,a]$$

for a in R^m. Let $\phi_{ij}(\lambda)$ denote the i-j^{th} element of $\Phi(\lambda)$. Then

$$\phi_{k\ell}(\lambda) \quad = \quad \int_{-\infty}^{\infty} e^{2\pi i(k-\ell)\Delta\gamma}\, p(\lambda,\gamma)\, d\gamma \; .$$

Now for any integer n:

$$\int_{-\infty}^{\infty} e^{2\pi in\Delta\gamma}\, p(\lambda,\gamma)\, d\gamma \;=\; \sum_{-\infty}^{\infty} \int_{\frac{-1}{2\Delta}}^{\frac{1}{2\Delta}} e^{2\pi in\Delta\gamma}\, p(\lambda,\, \gamma+\tfrac{k}{\Delta})\, d\gamma$$

$$= \int_{\frac{-1}{2\Delta}}^{\frac{1}{2\Delta}} e^{2\pi in\Delta\gamma} \left(\sum_{-\infty}^{\infty} p(\lambda,\, \gamma+\tfrac{k}{\Delta}) \right) d\gamma \qquad \text{a.e. in } \lambda.$$

Hence it follows that

$$[\Phi(\lambda)a,a] \;=\; \int_{-\infty}^{\infty} \left| \sum_{0}^{m-1} a_k\, e^{2\pi ik\Delta\gamma} \right|^2 p(\lambda,\gamma)\, d\gamma$$

$$\geq \; \mu(\lambda)[a,a]\;.$$

Hence

$$\log\, |\Phi(\lambda)| \;\geq\; \log\, \mu(\lambda)$$

and the result follows.

If (4.2) holds, then by the Rozanov-Helson-Lowdenslager factorization theorem [12], [13] we have

$$\Phi(\lambda) \;=\; \psi(\lambda)\; \psi(\lambda)^* \tag{4.5}$$

where

$$\psi(\lambda) \;=\; \int_0^{\infty} e^{2\pi i\lambda\sigma}\, W(\sigma)\, d\sigma \tag{4.6}$$

where

$$\int_0^{\infty} ||W(\sigma)||^2\, d\sigma \;<\; \infty \quad . \tag{4.7}$$

And as a consequence, we have the representation:

$$s(t,\cdot) \;=\; \int_0^{\infty} W(\sigma)\, N(t-\sigma)\, d\sigma \qquad -\infty < t < \infty \tag{4.8}$$

where $N(\cdot)$ is white noise in $L_2(-\infty,\infty)$. Note that (4.8) defines a Hilbert-Schmidt operator on $L_2(-\infty,\infty)$ into $L_2[0,T_1]$, for $T_1 < \infty$, by virtue of (4.7).

Let us list some examples of spectral desnities where (4.2) is satisfied.

(i) The spectral density

$$p_2(\lambda_1,\lambda_2) = \frac{1}{(a^2 + 4\pi^2\lambda_1^2 + 4\pi^2\lambda_2^2)^2} .$$

This field can be realized as the solution of the differential equation:

$$af(t_1,t_2) - \left(\frac{\partial^2}{\partial t_1^2} + \frac{\partial^2}{\partial t_2^2}\right) f(t_1,t_2) = N(t_1,t_2) ,$$

$$(t_1,t_2) \in R^2$$

where $N(t_1,t_2)$ is white noise in $L_2(R^2)$. Here $\left(\frac{\partial^2}{\partial t_1^2} + \frac{\partial^2}{\partial t_2^2}\right)$ denotes the closed linear operator with zero boundary conditions, with domain dense in $L_2(R^2)$.

(ii) The spectral density

$$p_2(\lambda_1,\lambda_2) = \frac{1}{(k^2 + 4\pi^2\lambda_1^2 + 4\pi^2\lambda_2^2)^{3/2}}$$

This is an isotropic random field with

$$E[s(t_1 + \tau, t_2 + \tau) s(t_1,t_2)] = \exp -k\tau .$$

(iii) The isotropic homogeneous random field with spectral density along any line (that is, spectral density of the process

$$f(t) = s(t_1 + t, t_2 + t))$$

given by

$$\frac{1 + a_1\lambda^2}{(1 + a_2\lambda^2)^{11/6}}$$

corresponding to the Von Karman model of turbulence spectral density.

(iv) The non-isotropic spectral density

$$p_2(\lambda_1,\lambda_2) = \frac{1}{(a^2 + 4\pi^2\lambda_1^2)} \exp - \frac{1}{2} b^2\lambda_2^2$$

where the second factor does not satisfy the factorizability condition.

We shall now indicate how a Kalman filter realization can be obtained under condition (4.3), or actually, the representation (4.8). We should note at this point that $\Phi(\lambda)$ is not necessarily a rational function of λ. It is not in fact in any of the examples (i) through (iv). Hence the state-space representation essential for developing the Kalman filtering theory, requires that the state-space be non-finite-dimensional, and we follow the technique devised by the author in [14].

Let $\mathcal{H}$ denote the L_2-space of $m \times 1$ matrix functions $f(t)$, $0 \leq t < \infty$, with norm defined by

$$||f(\cdot)||^2 = \int_0^\infty ||f(t)||^2 \, dt .$$

We can then define the linear bounded operator B mapping R^m into $\mathcal{H}$ by:

$$BN = f ; \quad f(t) = W(t)N , \quad 0 \leq t < \infty$$

where $W(\cdot)$ is the function defined by (4.6). Let $T(t)$ denote the shift semigroup over $\mathcal{H}$ defined by

$$T(t)f = g ; \quad g(s) = f(s+t) , \quad 0 \leq s \leq \infty .$$

Note that $T(t)BN$ is the function

$$W(t+s)N , \quad 0 \leq s < \infty .$$

Let A denote the infinitesimal generator of the semigroup $T(t)$. Define the operator C by

$$Cf = \lim_{\Delta \to 0} S_\Delta f$$

if the limit exists, where

$$S_\Delta f = \frac{1}{\Delta}\int_0^\Delta f(t)\,dt$$

and S_Δ maps $\mathcal{H}$ into R^m for each Δ. Thus defined,*/ the domain of C contains the class of continuous functions in $\mathcal{H}$ and thus has a dense domain, and is linear. It is not closed however, or even closeable. But, for any f in $\mathcal{H}$, clearly

$$T(t)f \in \text{domain of } C, \quad \text{a.e. } 0 \leq t < \infty ,$$

and, so does

$$\int_0^t T(t-\sigma)Bf(\sigma)\,d\sigma \quad ,$$

a.e. in $0 \leq t < \infty$. In fact

$$C\int_0^t T(t-\sigma)Bf(\sigma)\,d\sigma = \int_0^t W(t-\sigma)\,f(\sigma)\,d\sigma \ , \text{ a.e. } 0 \leq t < \infty$$

$$= \int_0^t CT(t-\sigma)Bf(\sigma)\,d\sigma, \text{ a.e. } 0 \leq t < \infty.$$

With the definitions we can now state

Theorem. Assume that the representation (4.8) holds. Then we have the state-space representation:

$$s(t,\cdot) = Cx(t) \qquad \text{a.e. } 0 \leq t < \infty, \tag{4.9}$$

$$\dot{x}(t) = Ax(t) + BN(t) \ , \quad \text{a.e. } t > 0 \ , \tag{4.10}$$

$$x(0) = \xi$$

where for each t, the state $x(t) \in \mathcal{H}$, and the solution of the equation (4.10) has to be interpreted in the generalized sense (see [11]), and ξ is a Gaussian random variable in $\mathcal{H}$ with covariance (operator)

$$\int_0^\infty W(\sigma)\,W(\sigma)^*\,d\sigma$$

*/ If the function W(t) were continuous in t, $0 \leq t < \infty$, then we could define $Cf = f(0)$, taking the domain of C to be space of functions continuous in $0 \leq t < \infty$. See [14].

and is independent of $N(\cdot)$, the latter being white noise in $\mathscr{H}$. Moreover, we have the Kalman-filter equations:

$$\dot{\hat{x}}(t) = A\hat{x}(t) + (CP(t))^*[v(t) - C\hat{x}(t)]$$

$$x(0) = 0$$

where $P(t)$ satisfies the Riccati equation:

$$[\dot{P}(t)x,x] = [P(t)x, A^*x] + [A^*x,P(t)x] + [Bx,Bx] - [(CP(t))^*x, (CP(t))^*x] \tag{4.11}$$

for $x \in \mathscr{D}(A^*)$, and $P(t)$ maps $\mathscr{H}$ into the domain of C; and finally

$$\hat{s}(t) = C\hat{x}(t)\,, \qquad \text{a.e. } 0 \le t$$

where

$$v(t) = \text{col. } v(t,k_\Delta), \qquad k = 0, 1, \ldots, (m-1)$$

and

$$\hat{s}(t) = E[s(t) \mid v(s), s \le t]\,.$$

Finally the likelihood-ratio formula for the discretized case can be written

$$\exp - \frac{1}{2}\left\{\int_0^{T_1} [\hat{s}(t),\hat{s}(t)]dt - 2\int_0^{T} [\hat{s}(t),v(t)]dt + \int_0^{T_1} \operatorname{Tr} C(CP(t))^*dt\right\}$$

<u>Proof</u>. The state-space representation (4.9) is immediate since solving (4.10) we have

$$x(t) = T(t)x(0) + \int_0^t T(t - \sigma)\, BN(\sigma)\, d\sigma$$

and

$$Cx(t) = CT(t)x(0) + \int_0^t W(t - \sigma)\, N(\sigma)\, d\sigma\,, \quad \text{a.e.}$$

and $x(0)$ is chosen so that

$$CT(t)x(0) = \int_{-\infty}^{0} W(t - \sigma)\, N(\sigma)\, d\sigma \quad , \quad \text{a.e.,} \quad 0 \leq t < \infty \, .$$

The proof of the Kalman filter equations follows generally the corresponding result for the case where C is bounded. The main step in the case where C is unbounded is to prove existence of solution to the Riccati equation (4.11). For the case where W(·) is continuous, a proof is given in [15]; the extension to the case where W(·) is not necessarily continuous will appear elsewhere.

REFERENCES

1. Nash, R.S., Jr., S.K. Jordan, Statistical Geodesy--An Engineering Perspective, Proceedings of the IEEE, 66 (1978), No. 5.

2. Rozanov, Y., On the Theory of Homogeneous Random Fields, Math' Sbornik (USSR), 32 (1977), 1-18.

3. Katani, S., Lecture Notes on Markov Random Fields, UCLA-ENG-7340 (1979).

4. Kallianpur, G., V. Mandrekav, The Markov Property for Generalized Gaussian Random Fields, Ann. Inst. Fourier Grenoble, 24 (1974), No. 2.

5. Larrimore, W.E., Statistical Inference on Stationary Random Fields, Proceedings of the IEEE, 65 (1977), No. 6.

6. Woods, J.W., C.H. Radewan, Kalman Filtering in Two Dimensions, Transactions Information Theory, 23 (1977), No. 4.

7. Balakrishnan, A.V., Likelihood Ratios for Signals in Additive White Noise, Lietuvos Matematikos Rinkinys, 18 (1978), No. 3.

8. Krein, M.G., I.C. Gohberg, Theory and Application of Volterra Operators in Hilbert Space, A.M.S. Translation (1970).

9. Gelfand, I.M., N.Ya. Vilenkin, Generalized Functions, Vol.4, Academic Press, New York (1964).

10. Balakrishnan, A.V., Non-Linear White Noise Theory, Multivariate Analysis, 5 (1980), 97-109.

11. Balakrishnan, A.V., Applied Functional Analysis, 2nd ed., Springer-Verlag (1981).

12. Helson, H., Lectures on Invariance Subspaces, Academic Press, New York (1964).

13. Rozanov, Y.A., Innovation Processes, Scripta Technica (1977).

14. A.V. Balakrishnan, Stochastic Filtering and Control: A General Theory, in Control Theory of Systems Governed by Partial Differential Eqautions, ed. Aziz et al., Academic Press (1977).

15. Balakrishnan, A.V., On a Class of Riccati Equations in a Hilbert Space, J. of Applied Math and Optimization (1980).

Research supported in part under grant no. 78-3550, Applied Mathematics Division, AFOSR, United States Air Force.

Lecture Notes in Control and Information Sciences

Edited by A. V. Balakrishnan and M. Thoma

Vol. 22: Optimization Techniques
Proceedings of the 9th IFIP Conference on
Optimization Techniques,
Warsaw, September 4–8, 1979
Part 1
Edited by K. Iracki, K. Malanowski, S. Walukiewicz
XVI, 569 pages. 1980

Vol. 23: Optimization Techniques
Proceedings of the 9th IFIP Conference on
Optimization Techniques,
Warsaw, September 4-8, 1979
Part 2
Edited by K. Iracki, K. Malanowski, S. Walukiewicz
XV, 621 pages. 1980

Vol. 24: Methods and Applications
in Adaptive Control
Proceedings of an International Symposium
Bochum, 1980
Edited by H. Unbehauen
VI, 309 pages. 1980

Vol. 25: Stochastic Differential Systems –
Filtering and Control
Proceedings of the IFIP-WG7/1 Working Conference
Vilnius, Lithuania, USSR, Aug. 28 – Sept. 2, 1978
Edited by B. Grigelionis
X, 362 pages. 1980

Vol. 26: D. L. Iglehart, G. S. Shedler
Regenerative Simulation of Response
Times in Networks of Queues
XII, 204 pages. 1980

Vol. 27: D. H. Jacobson, D. H. Martin, M. Pachter, T. Geveci
Extensions of Linear-Quadratic Control Theory
XI, 288 pages. 1980

Vol. 28: Analysis and Optimization of Systems
Proceedings of the Fourth International
Conference on Analysis and Optimization of Systems
Versailles, December 16–19, 1980
Edited by A. Bensoussan and J. L. Lions
XIV, 999 pages. 1980

Vol. 29: M. Vidyasagar,
Input-Output Analysis of Large-Scale
Interconnected Systems –
Decomposition, Well-Posedness and Stability
VI, 221 pages. 1981

Vol. 30: Optimization and Optimal Control
Proceedings of a Conference Held at
Oberwolfach, March 16–22, 1980
Edited by A. Auslender, W. Oettli, and J. Stoer
VIII, 254 pages. 1981

Vol. 31: Berc Rustem
Projection Methods in Constrained
Optimisation and Applications
to Optimal Policy Decisions
XV, 315 pages. 1981

Vol. 32: Tsuyoshi Matsuo,
Realization Theory of
Continuous-Time Dynamical Systems
VI, 329 pages, 1981

Vol. 33: Peter Dransfield
Hydraulic Control Systems –
Design and Analysis of Their Dynamics
VII, 227 pages, 1981

Vol. 34: H.W. Knobloch
Higher Order Necessary Conditions
in Optimal Control Theory
V, 173 pages, 1981

Vol. 35: Global Modelling
Proceedings of the IFIP-WG 7/1 Working
Conference Dubrovnik, Yugoslavia,
Sept. 1–5, 1980
Edited by S. Krčevinac
VIII, 232 pages, 1981

Vol. 36: Stochastic Differential Systems
Proceedings of the 3rd IFIP-WG 7/1
Working Conference
Visegrád, Hungary, Sept. 15–20, 1980
Edited by M. Arató, D. Vermes, A.V. Balakrishnan
VI, 238 pages, 1981